Studies in Applied Philosophy, Epistemology and Rational Ethics

Volume 24

About this Series

Studies in Applied Philosophy, Epistemology and Rational Ethics (SAPERE) publishes new developments and advances in all the fields of philosophy, epistemology, and ethics, bringing them together with a cluster of scientific disciplines and technological outcomes: from computer science to life sciences, from economics, law, and education to engineering, logic, and mathematics, from medicine to physics, human sciences, and politics. It aims at covering all the challenging philosophical and ethical themes of contemporary society, making them appropriately applicable to contemporary theoretical, methodological, and practical problems, impasses, controversies, and conflicts. The series includes monographs, lecture notes, selected contributions from specialized conferences and workshops as well as selected Ph.D. theses.

Advisory Board

More information about this series at http://www.springer.com/series/10087

Flavia Santoianni
Editor

The Concept of Time in Early Twentieth-Century Philosophy

A Philosophical Thematic Atlas

Springer

Editor
Flavia Santoianni
Department of Humanities
Section of Philosophy
University of Naples Federico II
Naples
Italy

ISSN 2192-6255 ISSN 2192-6263 (electronic)
Studies in Applied Philosophy, Epistemology and Rational Ethics
ISBN 978-3-319-24893-6 ISBN 978-3-319-24895-0 (eBook)
DOI 10.1007/978-3-319-24895-0

Library of Congress Control Number: 2015952525

Springer Cham Heidelberg New York Dordrecht London

Printed on acid-free paper

Springer International Publishing AG Switzerland is part of Springer Science+Business Media
(www.springer.com)

Acknowledgments

A special thanks goes to Lorenzo Magnani, who has constantly followed my research in these years, for his encouragement.

I am very grateful to Matthias Kaufmann, Louis Begioni, Guglielmo Tamburrini, and Akiko Frischhut for their precious interpretation to discuss the perspectives on *time* expressed by this volume.

I want to thank all authors who, with their significative contributions, have made the publication of this volume possible; in particular, Michele Malatesta and Giuseppe Cacciatore for their extraordinary contribution.

I am so grateful to Rocco Pititto, Nicola grana, and Rosario Diana for their continuous support and guidance and to all young researchers of our Department for their great effort, which I deeply appreciated—especially to Roberto Evangelista, who has dedicated so much of his time for editing the bibliography of this volume.

Finally, my last but not least thanks go to Leontina Di Cecco for her constant endorsement.

Contents

Introduction 1
Rosario Diana

Spaces of Thinking 5
Flavia Santoianni

Space in Education 15
Monica Sorrentino

Part I Phenomenology and Perception of Time

Introduction 29
Alessandro Arienzo

Phenomenology and Perception of Time Maps 35
Flavia Santoianni

Time and Reality in the Thought of Henri Bergson 39
Mirko Di Bernardo

Concepts of Time in Husserl 59
Felice Masi

L'évasion de l'être. Jean-Paul Sartre and the Phenomenology of Temporality 77
Armando Mascolo

The Time of the Body in Maurice Merleau-Ponty 85
Maria Teresa Catena

Time Out of Joint: Gilles Deleuze and Felix Guattari on Time and Capitalism 95
Alessandro Arienzo

On Derrida's Critique of the Metaphysics of Presence. Implications for Scientific Inquiry 103
Marco Stimolo

Commentary: The Phenomenology and Perception of Time 109
Akiko M. Frischhut

Part II Language and Thinking of Time

Introduction 121
Roberto Evangelista

Language and Thinking of Time Maps 125
Flavia Santoianni

Historiographical Language and Temporality in Ernst Cassirer 129
Fabrizio Lomonaco

Wittgenstein on Time: From the Living Present to the Clock Time 137
Giorgio Rizzo

Sign(s) of the Time: Time and Understanding in Heidegger's Phenomenological–Ontological Hermeneutics 149
Simona Venezia

Psychoanalysis at the Test of Time: Jacques Lacan's Teaching 157
Marco Castagna

Time, Narration, Memory: Paul Ricoeur's Theory of History 167
Giuseppe Cacciatore

Historical Heterochronies: Evenemential Time and Epistemic Time in Michel Foucault 175
Agostino Cera

Commentary: Talking About Time and Whether We Should Measure It 183
M. Kaufmann

Part III Science and Logic of Time

Introduction 189
Mirko Di Bernardo

Note of Introduction 197
Gianluca Giannini

Science and Logic of Time Maps 199
Flavia Santoianni

The Linguistics of the 1900s from Ferdinand de Saussure to Gustave Guillaume Between Synchrony and Diachrony 203
Rocco Pititto

Time and Relativity of Time in Einstein's Theory of Special Relativity 211
Salvatore Principe

Tenses and Temporality in Reichenbach's Thought 217
Michele Malatesta

The Concept of Time in Prigogine 229
Nicola Grana

The End of Time: New Perspectives of Self-identification for Man 239
Gianluca Giannini

Commentary: Gustave Guillaume Between Linguistics and Philosophy of Language: A New Point of View 247
Louis Begioni

Commentary: Einstein, Prigogine, Barbour, and Their Philosophical Refractions 249
Lorenzo Magnani

Commentary: Reichenbach's Verbal Tenses in the Context of Discovery About Computing Systems 253
Guglielmo Tamburrini

Editor Biography 259

Introduction

Rosario Diana

For better or for worse, it is well known that the various philosophies of our Western tradition—the overtly systematic ones and those which did not manage or did not wish to be so—have been skillful and more or less successful attempts to organize, in a broad sense, the world and its vast and meaningful aspects. With respect to these, an atlas is a historical–theoretical product of a metaphysical nature: it is an instrument which—starting from an ordering principle—aims to offer the reader a "systematization" of those cultural productions that can be gathered under the concept of "philosophy" (or close to the latter in content and style of thought, though originating in other scientific contexts) and which contain a reading of the real. In brief: if, in general, the philosophies wish to furnish reference points by which to orient ourselves in our life in the world, an atlas aims to be a kind of map, a "road map" (Merker 2002: 11) to orient ourselves among these various orientative proposals.

This work edited by Flavia Santoianni is not the first, and will not be the last philosophical atlas[1], but has at least three characteristics that make it particularly trustworthy and of considerable importance. The first refers to the makeup of its various parts, entrusted to talented and authoritative scholars, specialists in the authors and problems that are treated by the various articles. In the age of professionalism in philosophy (Marconi 2014; Cacciatore 2015; Diana 2015), this fact may seem controversial to some. I believe, however, that the reader will be happy with it, because specialization always represents a kind of guarantee and legitimates one in supposing that the scholar who writes has in-depth and longstanding competence. The second regards the conciseness of the individual contributions, which in a few pages, with clarity and richness of primary and secondary documentation,

[1]For an initial panorama of the philosophical atlases published in Italian, see Diana 2013.

R. Diana (✉)
National Research Council, Institute for the History of Modern Philosophical and Scientific Thought, via Porta di Massa 1, 80133 Naples, Italy
e-mail: rosariodiana@teletu.it

F. Santoianni (ed.), *The Concept of Time in Early Twentieth-Century Philosophy*, Studies in Applied Philosophy, Epistemology and Rational Ethics 24,
DOI 10.1007/978-3-319-24895-0_1

offer the reader a first contact with the topic addressed: an exhaustive contact, if the reader is seeking only to acquire correct information about the topic; a stimulating and orienting contact for those who are approaching the subject for the first time to then further research the questions examined in a more in-depth manner. All this makes this book a very valuable traveling companion.

The third characteristic merits a more ample discussion. There are geographical maps in the atlas, which situate the currents and authors treated in their various geo-cultural regions of origin. But the book also contains conceptual maps, whose particular function can be understood only by referring to the preceding works of the editor (Santoianni 2011, 2014), of which what we are presenting here is the applicative precipitate. These maps are not the mere and simplifying graphic transposition of the contents conveyed in the contributions presented, but are—as we shall see—the result of a very well-founded and well-reasoned theoretical approach.

The point of departure is the "vision of a situated and embodied mind" (Santoianni 2014: 38): a relationship between mind and body that can be read in the light of the musical concept of "orchestration" (Cingolani and Metta 2015: 29). As the composer constructs his or her own musical thought and creates sounds and their sequences in relation to the orchestral instruments he or she intends to use, which from the beginning—with their organological, timbric, and technical characteristics—are present in the compositional process and guide the artistic creation, so

> the mind-body relationship is not that between controller (brain) and controlled (body), but is a much more complex dynamic in which the controller develops control strategies, which are possible only because it has at its disposal the morphological characteristics of the controlled entity (Cingolani and Metta 2015: 29).

This means that even among the most elementary processes of knowledge, that is, those agents at the level of the *implicit*—which refers to those abilities that cannot be translated into a list of instructions (walking, riding a bicycle, catching a flying ball)—and those more complex and sophisticated ones, that is, those connected to the dimension of the *explicit*—which includes abilities that can be broken down linguistically into a series of algorithms, for instance: making origami or solving an equation—(Carr 2015: 23–24), there must be a relation. Knowledge cannot but derive from complicity between the *implicit* and the *explicit*. Flavia Santoianni writes:

> Implicit elaboration can […] serve as a cognitive antecedent with adaptive functions in the primary evolutive phases of cognitive development, in philogenesis as well as in ontogenesis. This can occur in every kind of learning situation, both in the first phases of conceptual development and as a primitive form—though one constantly present in the course of individual development—of cognitive organization. […] In the possible collaborative relation between explicit and implicit, continuous bi-directional transits would be activated in a flexible manner from the implicit to the explicit and vice versa (Santoianni 2014: 49, 51).

However, it is necessary to identify a middle term between the implicit and the explicit that would favor reciprocal exchange between them. Santoianni holds that the "unity of the linkage between the two levels", that is to say the "interface" between them, can be recognized in the "elementary logics of thought" (Santoianni 2014: 48, 78), which are articulated in the two macro-classes of *sequence* and *parallelism*. The first of these (*sequence*) branches out into the classes: *union* (relation: *integration* and *sequentiality*) and *separation* (relation: *individuation*). The second (*parallelism*) has below it the classes: *separation* (relation: *comparison*) and *correlation* (relation: *inference* and *correlation*). That which I have been able to refer to here only with extreme brevity, is instead discussed in the work of Flavia Santoianni in detail and with due reference to the existing scientific literature on the topic.

An adequate form of expression of the classes and relations of the elementary logics is found—according to Santoianni—in the graphic representations proper to the so-called "spatial thought". Schemas and diagrams are presented and discussed in Flavia Santoianni's previous work in connection with the individual classes and relations proper to the elementary logics of thought. In the latter, one reads in the book,

> the role of graphic representation in the formation of thought is not *successive* to explicit analysis; in other words, it is not used to explain better and to make more deeply understood that which has already been explained in words. The implicit dimension in this theory carries out [...] a collaborative role; therefore, its usage through spatial representations *precedes*, and does not follow, verbalization, or flanks it (Santoianni 2014: 86).

Behold, therefore, the explanation of the third characteristic of this atlas: that which constitutes its true specificity, in line with the previous scientific work of Flavia Santoianni. The role of conceptual maps must not be understood as the concise reduction of the contributions, or as a mnemonic device. The individual windows contain whole passages taken from the collected articles, and the graphic schemas, taken jointly, correspond to the classes and relations of the elementary logics active in the thought processes implemented by the contributors. In substance, the maps reproduce in "spatial" form the conceptual directionalities that support the discourses and arguments. According to Santoianni's indications in the passage above, which we have quoted in its entirety, such maps should be read through and meditated on either before or simultaneous to the reading of the various chapters: and this suggested style of learning means to solicit precisely that collaboration between implicit and explicit that is the basis for complete and solid knowledge, in which the philogenetic heritage and ontogenetic identification interweave.

We stand before a pedagogical perspective of great breadth, which flees from every abstract conception of the mind—linking it to its (too often forgotten) ancestral and corporeal roots—and conceives the learning process as the result of a reciprocal relation between the unconscious/nonverbal element and the conscious/linguistic element.

References

Cacciatore, G. (2015). Del "pensare in proprio" nell'epoca delle filosofie mediatiche. *Research Trends in Humanities II, 2*, 33–39.

Carr, N. (2015). *La gabbia di vetro. Prigionieri dell'automazione* (2014). (S. Garassini & G. RomanoItalian, Italian Trans.). Milano: Raffaello Cortina.

Cingolani, R-M., Metta, G. (2015). *Umani e umanoidi. Vivere con i robot*. Bologna: il Mulino.

Diana, R. (2013). Cartografia e filosofia interculturale nell'*Atlante* di Holenstein. In Id., *Identità individuale e relazione intersoggettiva. Saggi di filosofia interculturale*, 99–114. Roma: Aracne.

Diana, R. (2015). Ricerca filosofica e scelta etica per la sua disseminazione. *Research Trends in Humanities II, 2*, 40–47.

Marconi, D. (2014). *Il mestiere di pensare*. Turin: Einaudi.

Merker, N. (2002). *Atlante storico della filosofia*. Roma: Editori Riuniti.

Santoianni, F. (2011). Educational models of knowledge prototypes development. *Mind & Society, 10*, 103–129.

Santoianni, F. (2014). *Modelli di studio*. Erickson: Apprendere con la teoria delle logiche elementari. Trento.

Spaces of Thinking

Flavia Santoianni

> *Ease of recognition may be strongly affected by what information is explicit in a representation, and what is only implicit.*
>
> Larkin and Simon 1987
>
> *It is especially important to determine how the symbol systems of maps and diagrams interact with pre-attentive psychological processes... A person cannot therefore willfully influence them.*
>
> Winn 1991
>
> *The pre-attentive visual processes draw from cognitive resources that do not interfere with those required for attentional processing.*
>
> Nesbit and Adesope 2006

Abstract Spatial representations have been considered for their high didactical efficacy, as *concept maps* and *mind maps*. Graphical and spatial representations may be seen as key elements of knowledge management and may contribute to enhancing spatial knowledge. Even if isomorphisms between the physical and the mental dimension can be controversial in the field of spatial knowledge, it is nevertheless interesting to study the role of spatial interpretation in knowledge management processes. In science education, spatial skills are actually highly required due to the development of new technologies and their highly demanding spatial tasks they often work intertwined with other abilities, such as logical reasoning and verbal skills. Despite spatial knowledge being significative in the overall field of learning management, there is instead little research that shows its complex role in the comprehension of concepts in the humanistic fields, e.g. in philosophical conceptual reasoning. The *Elementary Logic Theory*—which underlies the philosophical maps of this Atlas—identifies at the basis of complex thinking prototypical knowledge units that may be activated in the possible collaboration between explicit and implicit thinking. In *EL Theory*, spatiality is considered the most suitable transition format for the hypothesized collaboration between implicit and explicit processing because it shares common aspects with both of them. The focus of *EL Theory* is to research particularly the role of the implicit in knowledge comprehension through spatial representations. In this Atlas, the role of *Elementary*

F. Santoianni (✉)
Philosophy Section, Department of Humanities, University of Naples Federico II,
via Porta di Massa 1, 80133 Naples, Italy
e-mail: bes@unina.it

F. Santoianni (ed.), *The Concept of Time in Early Twentieth-Century Philosophy*,
Studies in Applied Philosophy, Epistemology and Rational Ethics 24,
DOI 10.1007/978-3-319-24895-0_2

Logic Theory is highlighted in shaping philosophical knowledge and comprehension of philosophical concepts and their intertwining.

Scientific areas of research usually refer to graphical representations to deepen the understanding of scientific phenomena (Larkin and Simon 1987; Miller 1996; Mohler 2008); in particular, attention has been focused on graphs (Roth et al. 1999). Spatial representations have also been considered for their high didactical efficacy, as *concept maps* (Novak 1998; Novak and Gowin 1984) and *mind maps* (Buzan and Buzan 2000). The graphical representation of knowledge through maps has been described as a way to express meanings inherent in the material to be learned according to its property of explicitly expressing relations between ideas, theories, and concepts.

Graphical and spatial representations may be seen as key elements of knowledge management and may contribute to enhancing spatial knowledge, which in its most elemental type may be considered the knowledge of the identities and appearances of objects or environmental features (McNamara et al. 2008). In this last sense, *landmark knowledge* is recognized as the first building block of other spatial knowledge (Siegel and White 1975). Sequences of landmarks are represented by *route knowledge*, which consists of discrete chunks of information representing sequential locations (Hirtle and Hudson 1991) allowing one to get from one point to another (Taylor and Tversky 1992). Procedural descriptions are in fact related to the sequential records of the space involved.

Survey knowledge may instead give an overall picture of a context as an aid in orientation, resembling a 'map in the head' (Kuipers 1982) which locates objects in a general frame of reference (Hirtle and Hudson 1991). It gives a *configural*[1] holistic representation about the location of objects (Taylor and Tversky 1992). The overall configuration of an environment and its definition within a common reference system is regulated by survey knowledge, the most sophisticated type of spatial knowledge, which is also referred to as a cognitive map (Tolman 1948). Even if isomorphisms between the physical and the mental dimensions can be controversial in the field of spatial knowledge (McNamara et al. 2008), it is nevertheless interesting to study the role of spatial interpretation in knowledge management processes. Just as a geographical map is used to orient oneself in a territory, a concept/mind map may be considered a tool to interpret and process knowledge, displaying links between concepts and developing paths of reasoning (Gineprini and Guastavigna 2002).

In science education, spatial skills—the capacity to understand and remember the spatial relations among objects—are enhanced to construct, from verbal descriptions in textbooks, mental models of objects which may sustain the mastering of the subject matter. More in general, spatial skills are actually highly required due to the development of new technologies and their highly demanding

[1]The terms featural and configural may be used, respectively, to refer to procedural/route-like and holistic/map-like representations.

spatial tasks; they often work intertwined with other abilities, such as logical reasoning and verbal skills.

Despite spatial knowledge being significative in the overall field of learning management and the fact that it is deeply studied in its several aspects in relation to scientific areas and STEM fields (Metz et al. 2012), there is instead little research that shows its complex role in the comprehension of concepts in the humanistic fields, apart from its general didactical sense—e.g. in philosophical conceptual reasoning (Santoianni 2014a). Moreover, the improvement of spatial knowledge is related to environmental factors, the manipulation of which through educational experience may increase learning performances. This aspect implies an educational interest towards this research topic and makes it attractive for experimental research in the human sciences.

A key related point concerns the relationships between spatial knowledge and language; for instance, while graphs are mainly considered as topological, verbal linguistic representations may be seen as typological (Bastide 1990), due to their categorizing character. The graphical and spatial representations that combine topological and typological features should be more effective, as they represent multimodal expressions —as in Paivio's *Dual Coding Theory* (Paivio 1986)—even if the phenomenon known as *verbal overshadowing*[2] has to be considered (Fiore and Schooler 2002).

In concept/mind maps—which can be seen as an example of topological spatial representations—graphical representation of concepts are connected by lines that make explicit the internal relationship between these concepts within a geometric shape, through linking words that may express relations of sequentiality (and comparison), and inference. In the field of spatial knowledge, scholars have recently focused on environmental shape and its role in learning processes. Learners are indeed sensitive to contextual geometry in new environments (Shelton and McNamara 2001) which is an aid to orient oneself in them (Schmidt and Lee 2006; Hartley et al. 2004; Hermer and Spelke 1994).

Instead, in the *Elementary Logic Theory* (Santoianni 2011, 2014b)—which underlies the philosophical maps of this Atlas—the foreseen relations between concepts are more than the ones envisaged by conceptual and mind maps. *EL Theory* model concerns the *spaces of thinking* represented by elementary logics, to be understood as the knowledge prototypes that express the logical relationships underlying the organization of concepts in language and mathematics, and probably in other fields of knowledge, both theoretical and practical.

According to *EL Theory*, explicit knowledge is developed in relation to implicit functions of thinking, which may be interpreted as cognitive antecedents constantly active in cognitive processing, both in the early stages of conceptual development and as primitive forms of cognitive organization. Implicit knowledge may operate as a default level to be activated on demand regarding cognitive tasks in an individual's lifespan.

---

[2]This overshadowing is caused by verbalization and is evident in various forms of cognition; it consists of an impairment of *configural* processes in favour of *featural* representations.

Table 1 Elementary logics

ADD	Integration	Putting together two or more conceptual elements or addition of a conceptual element to a set of pre-existing elements
CHAIN	Sequentiality	Concatenation of two or more conceptual elements that are correlated under the relationship of the consequential links between them
EACH	Individuation	Focusing on one or more conceptual elements within a set of elements, distinguishing it/them from others
COMPARE	Comparison	Comparing two or more conceptual elements and the recognition of the possible similarities and differences between them
FOCUS	Inference	Deductive and inductive process according to which a conceptual element can be in a derivative relation with others
LINK	Correlation	Non-systematic identification of possible links between more conceptual elements related by connection points with a specific bridge function

Implicit functions play the role of prototypes of the possible models of explicit knowledge and are involved in the continuous cognitive collaboration with it through dynamic patterns as modes of connection between the two levels. Logical functions that may regulate these dynamic patterns are represented by elementary logics of thinking and by their spatial representations. Elementary logics underlie linguistic and mathematical thinking and may be expressed by spatial representations. In *EL Theory*, elementary logics are systematized in three classes—*union*, *separation*, and *correlation*—each subdivided in two functions: *integration* and *sequentiality* (union), *individuation* and *comparison* (separation), *inference* and *correlation* (correlation). Everyone learns through a combination of elementary logics (Table 1).

Indeed *EL Theory* identifies at the basis of complex thinking prototypical knowledge units that may be activated in the possible collaboration[3] between explicit and implicit thinking. Linking words and word tags—characteristic of the spatial structure of maps in traditional literature (see Sorrentino in this volume)—are substituted by logical and mathematical symbols (Table 2). The significance of graphical representations in the spatial comprehension of knowledge concepts is here studied mainly in its implicit dimension, without necessarily referring to its possible explicit expression.

In *EL Theory*, spatiality is considered the most suitable transition format for the hypothesized collaboration between implicit and explicit processings because it

[3]EL Theory proposes the hypothesis that the function performed by implicit learning may be integrative with regard to the explicit, thus overcoming the *polarity fallacy* (Reber 1993)—this is a crucial node of discussion in literature (Santoianni 2014a, b). In this case it could be *on demand*, to be activated at the request of the explicit in respect to the specificity of any cognitive task or to environmental requirements. Its presence in the cognitive system would be justified by its functional role of adaptive nature in philogenesis as in ontogenesis.

Table 2 Elementary logics symbols

ADD	*Integration*			
CHAIN	*Sequentiality*			
EACH	*Individuation*			
COMPARE	*Comparison*			
FOCUS	*Inference*			
LINK	*Correlation*	K		

shares common aspects with both of them (Lleras and Von Mühlenen 2004; Haun et al. 2005) (Table 3).

Even if both implicit and explicit may contribute to the cognitive activity processes, the focus of *EL Theory* is to research particularly the role of the implicit in knowledge comprehension through spatial representations.

However, spatiality of visual processing can be explicit or implicit. At the phylogenetic level, knowledge prototypes may have been characterized by only partially aware functions of recognition and identification (Ramachandran 2004) and by implicit spatial maps with an adaptive role which have anticipated the explicit linguistic thinking (Siegler et al. 1996). At the ontogenetic level, brain regions involved in visuospatial tasks seem to be correlated to automatic modes of thinking, while brain regions close to language centres appear to regulate the explicit expression of thinking, more related to the verbal sphere (Smith and DeCoster 2000). The relationship between visual processing and implicit processing is thus a characteristic of visuospatial representations (Chun and Jiang 1998). In the cognitive field, the nexus of the spatial representations of the schematic type (diagrams) with explicit processing has been recognized.

Visual processing appears to be one of the most significant aspects in spatial thinking development, which implies engendering internal non-linguistic mental images (Gardner 1983). Both explicit and implicit learning can be configured in internal images with a peculiar structure and definition in space (Entwistle and Smith 2002; Hollands and Spence 2001), and in mental models (Morra 2001) which can be stored in specifically spatial representational formats (Robinson et al. 1999).

Spatial representations have been traditionally distinguished from textual representations because the latter may contain implicit information about the logical

Table 3 Elementary logics and spatial representations

IMPLICIT ELEMENTARY LOGICS				IMPLICIT AND EXPLICIT SPATIAL REPRESENTATIONS
SEQUENCE	UNION	ADD	*Integration*	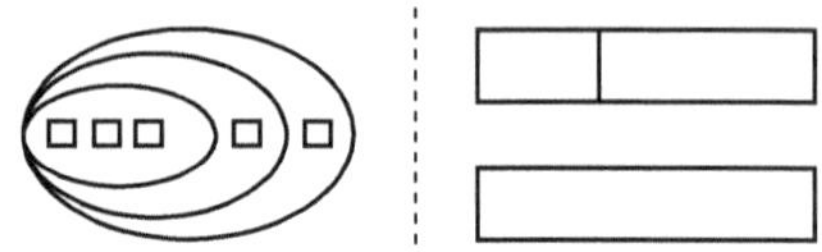
		CHAIN	*Sequentiality*	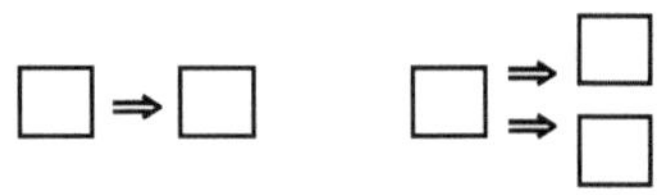
	SEPARATION	EACH	*Individuation*	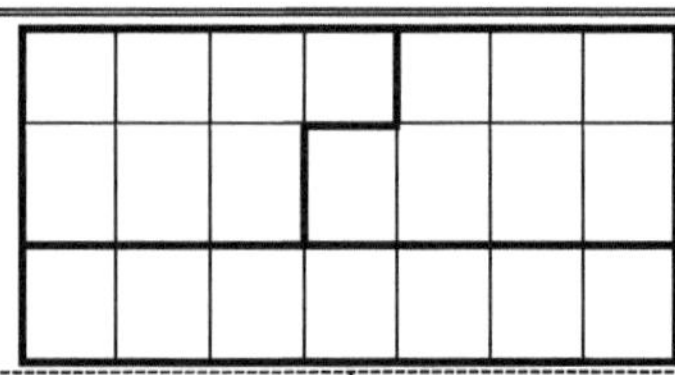
PARALLELISM		COMPARE	*Comparison*	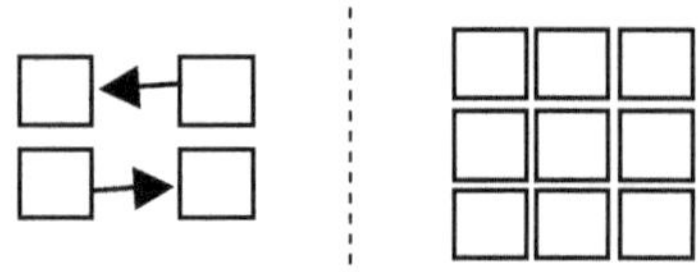
	CORRELATION	FOCUS	*Inference*	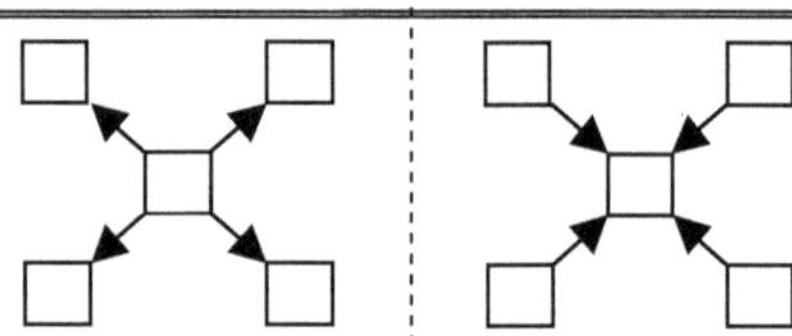
		LINK	*Correlation*	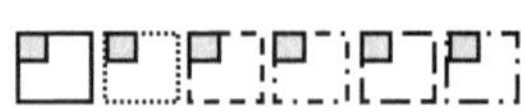

relationships between the parts of a text, while the spatial arrangement of textual information implies the idea that these relationships can be made explicit. As Larkin and Simon wrote in their famous article of 1987, diagrammatic representations are distinguishable from sentential representations because they can make explicit the implicit relationships possibly hidden between the different parts of a text (Larkin and Simon 1987). Spatial diagrammatic representations have been defined as *visual organizers* of thinking patterns which reflect the structure of knowledge contents (Clarke 1991) and play a significant role in learning management. The effectiveness of schematic representations (*graphic organizers*) lies therefore in the learner's possibility to explicitly identify the logical relations between parts of a text, in a more or less direct connection with the subsequent linguistic traceability characteristic of explicit learning (Santoianni 2006).

Graphical and spatial representations may be a learning facilitator (Nesbit and Adesope 2006). However, in some cases—such as the interpretation of unfamiliar graphs by scientists—the understanding of the meaning of a graph may be not self-evident as occurs with familiar graphs. It would instead be the result of a complex inferential process (Roth and Bowen 2003). The lack of familiarity with a specific representation, its contextual signs and inherent conventions, may be then an impediment to spatial encoding processes. So the cognitive system must make a verbal effort to elaborate and explain graphical representations.

According to this view, familiarity with a graphical representation may be due to its adaptive use and, in this sense, it can imply the involvement of the implicit processing. Following an evolutive model of implicit learning, implicit processing could originate from phylogenesis as the oldest form of learning—a process 'at the root' of the behavioural adaptive repertoire of every complex organism—appearing prior to the human race's explicit linguistic manner of communication (Reber 1992, 1993). Implicit learning could be a form of prototypical[4] cognitive processing (Santoianni 2011), linked to spatiality (Huttenlocher et al. 1991; Crawford et al. 2000), which can be used adaptively even in ontogenesis to represent the structure of the environment to solve problems and to produce basic—that can also be abstract—knowledge (Reber 1989).

In adaptive learning environments (Santoianni 2007), there may be predefined maps to provide adaptive learning guidance for learners (Shian-Shyong et al. 2007). To be effective, these maps should be built according to scientific models, to simplify and clarify abstract concepts, and to explain and understand theories (Dori and Barak 2001). Models can contribute to visualizing complex ideas, processes and systems. But what kind of scientific model could be used to organize maps in the humanities?

In this Atlas, the role of *Elementary Logic Theory* (Santoianni 2011, 2014b) is highlighted in shaping philosophical knowledge and comprehension of philosophical concepts and their intertwining. The use of graphical and spatial representations in the humanities—a field domain strictly dominated by linguistic expression—may be justified precisely by the role that adaptive processes play in any learning situation (Santoianni 2007). And, if the familiarity with a graphical representation is enough for a cognitive system to read it without the aid of linking or tag words, the continuative use of graphical and spatial representations in a humanistic field, like philosophy, may enhance philosophical and spatial mastery through implicitly linked map-like representations, such as *EL maps*.

---

[4]See *knowledge prototypes* (Lambiotte et al. 1989; O'Donnell et al. 2002).

References

Bastide, F. (1990). The iconography of scientific texts: Principles of analysis. In M. Lynch & S. Woolgar (Eds.), *Representation in scientific practice* (pp. 187–229). Cambridge: MIT Press.

Buzan, B., & Buzan, T. (2000). *The mind map book.* London: BBC Books.

Chun, M., & Jiang, Y. (1998). Contextual cueing: implicit learning and memory of visual context guides spatial attention. *Cognitive Psychology, 36*, 28–71.

Clarke, J. (1991). Using visual organizers to focus on thinking. *Journal of Reading, 34*(7), 526–534.

Crawford, E., Regier, T., & Huttenlocher, J. (2000). Linguistic and non-linguistic spatial categorization. *Cognition, 75*(3), 209–235.

Dori, J., & Barak, M. (2001). Virtual and physical molecular modeling: fostering model perception and spatial understanding. *Educational Technology & Society, 4*(1), 61–74.

Entwistle, N., & Smith, C. (2002). Personal understanding and target understanding: Mapping influences on the outcomes of learning. *British Journal of Educational Psychology, 72*, 321–342.

Fiore, S., & Schooler, J. (2002). How did you get here from there? verbal overshadowing of spatial mental models. *Applied Cognitive Psychology, 16*, 897–910.

Gardner, H. (1983). *Frames of mind. The theory of multiple intelligences.* Milano: Feltrinelli.

Gineprini, M., & Guastavigna, M (2002). Mappe, complessità, strutture di comprensione. *Insegnare* 12. http://www.pavonerisorse.to.it/cacrt/mappe/.

Hartley, T., Iris, T., & Neil, B. (2004). Geometric determinants of human spatial memory. *Cognition, 94*(1), 39–75.

Haun, D., Allen, G., & Wedell, D. (2005). Bias in spatial memory: a categorical endorsement. *Acta Psychologica, 118*, 149–170.

Hermer, L., & Spelke, E. (1994). A geometric process for spatial reorientation in young children. *Nature, 370*(6484), 57–59.

Hirtle, S., & Hudson, J. (1991). Acquisition of spatial knowledge for routes. *Journal of Environmental Psychology, 11*, 335–345.

Hollands, J., & Spence, I. (2001). The discrimination of grafical elements. *Applied Cognitive Psychology, 15*, 413–431.

Huttenlocher, J., Larry, H., & Duncan, S. (1991). Categories and particulars: Prototype effects in estimating spatial location. *Psychological Review, 98*(3), 352–376.

Kuipers, B. (1982). The'map in the head'metaphor. *Environment and Behavior, 14*, 202–220.

Lambiotte, J.G., Dansereau, D.F., Cross, D.R. & Reynolds, S.B. (1989). Multirelational semantic maps. *Educational Psychology Review*, 1:331-367.

Larkin, J., & Simon, H. (1987). Why a diagram is (Sometimes) worth ten thousand words. *Cognitive Science, 11*, 65–99.

Lleras, A., & Von Mühlenen, A. (2004). Spatial context and top-down strategies in visual search. *Spatial Vision, 17*(4–5), 465–482.

McNamara, T., Sluzenski, J., & Rump, B. (2008). Human spatial memory and navigation. In J. Byrne (Ed.) *Learning and memory: A comprehensive reference*, (pp. 157–178).

Metz, S., Donohue, S., & Moore, C. (2012). Spatial skills: A focus on gender and engineering. In B. Bogue & E. Cady (Eds.), *Apply research to practice (ARP) resources* (pp. 9–55). New York: Academic Press.

Miller, C. (1996). A historical review of applied and theoretical spatial visualization publications in engineering graphics. *Engineering Design Graphics Journal, 60*(3), 12–33.

Mohler, J. (2008). A review of spatial ability research. *Engineering Design Graphics Journal, 72* (3), 19–30.

Morra, S. (2001). On the information-processing demands of spatial reasoning. *Thinking & Reasoning, 7*(4), 347–366.

Nesbit, J., & Adesope, O. (2006). Learning with concept and knowledge maps: A meta-analysis. *Review of Educational Research, 76*(3), 413–448.

Novak, J. (1998). *Learning, creating, and using knowledge: Concept maps as facilitative tools for schools and corporations*. Mahwah: Lawrence Erlbaum.
Novak, J., & Gowin, B. (1984). *Learning how to learn*. Cambridge: Cambridge University Press.
O'Donnell, A.M., Dansereau, D.F. & Hall, R.H. (2002). Knowledge Maps as Scaffolds for Cognitive Processing.*Educational Psychology Review*, *14*(1):71–86.
Paivio, A. (1986). *Mental representations*. New York: Oxford University Press.
Ramachandran, V. (2004). *Che cosa sappiamo della mente*. Milano: Mondadori.
Reber, A. (1989). Implicit learning and tacit knowledge. *Journal of Experimental Psychology: General, 118*(3), 219–235.
Reber, A. (1992). The cognitive unconscious: An evolutionary perspective. *Consciousness and Cognition, 1*, 93–113.
Reber, A. (1993). *Implicit learning and tacit knowledge. An essay on the cognitive unconscious*. Oxford: Oxford University Press.
Robinson, D., Robinson, S., & Katayama, Andrew. (1999). When words are represented in memory like pictures: Evidence for spatial encoding of study materials. *Contemporary Educational Psychology, 24*(1), 38–54.
Roth, W.-M., & Bowen, M. (2003). When are graphs worth ten thousand words? *An Expert-Expert Study. Cognition and Instruction, 21*(4), 429–473.
Roth, W.-M., Bowen, M., & McGinn, M. (1999). Differences in graph related practices between high school biology textbooks and scientific ecology journals. *Journal of Research in Science Teaching, 36*, 977–1019.
Santoianni, F. (2006). *Educabilità cognitiva*. Roma: Carocci.
Santoianni, F. (2007). Bioeducational perspectives on adaptive learning environments. In F. Santoianni & C. Sabatano (Eds.), *Brain Development in Learning Environments. Embodied and Perceptual Advancements* (pp. 83–96). Cambridge: Cambridge Scholars Publishing.
Santoianni, F. (2011). Educational models of knowledge prototypes development. *Mind & Society, 10*, 103–129.
Santoianni, F. (Ed.). (2014a). *La filosofia nello spazio del pensiero*. Roma: Carocci.
Santoianni, F. (2014b). *Modelli di studio. Apprendere con la teoria delle logiche elementari*. Trento: Erickson.
Schmidt, T., & Lee Eun, Y. (2006). Spatial memory organized by environmental geometry. *Spatial Cognition and Computation, 6*(4), 347–369.
Shelton, A., & McNamara, T. (2001). Systems of spatial reference in human memory. *Cognitive Psychology, 43*(4), 274–310.
Shian-Shyong, T., Pei-Chi, S., Jun-Ming, S., Jui-Feng, W., & Wen-Nung, T. (2007). A new approach for constructing the concept map. *Computers & Education, 49*, 691–707.
Siegel, A., & White, S. (1975). The development of spatial representations of large-scale environments. In H. Reese (Ed.) *Advances in Child Development and Behavior* (Vol 10, pp. 9–55).
Siegler, R., Adolph, K., & Lemaire, P. (1996). Strategy choices across the life span. In Lynne Reder (Ed.), *Implicit learning and metacognition* (pp. 79–121). Mahwah: Lawrence Erlbaum.
Smith, E., & De Coster, J. (2000). Dual-Process models in social and cognitive psychology: Conceptual integration and links to underlying memory systems. *Personality and Social Psychology Review, 4*, 108–131.
Taylor, H., & Tversky, B. (1992). Spatial mental models derived from survey and route descriptions. *Journal of Memory and Language, 31*, 261–292.
Tolman, E. (1948). Cognitive maps in rats and men. *Psychological Review, 55*(4), 189–208.
Winn, W. (1991). Learning from maps and diagrams. *Educational Psychology Review*, *3*(3):211–247.

Space in Education

Monica Sorrentino

Abstract There is a varicty of way to organize and manage spatially knowledge in literature. In this present inquire only a few type relating to the context of education will be analyzed: two of the most widespread and oft-used in the field of didactic pedagogy and one that has recently been scientifically affirmed—though its study dates back twenty years. The parallel—and sometimes intertwining—analysis of these three interpretive perspectives could serve as an epistemological scaffolding allowing us to look out on a holistic experimental paradigm on the topic of spatialization in/of cognitive processes

In literature there is a variety of ways to represent knowledge—in the sense of organizing and managing it spatially. In this present inquiry only a few types relating to the context of education will be analyzed: two of the most widespread and oft-used in the field of didactic pedagogy and one that has recently been scientifically affirmed—though its study dates back 20 years. The parallel—and sometimes intertwining—analysis of these three interpretive perspectives could serve as an epistemological scaffolding allowing us to look out on a holistic experimental paradigm on the topic of spatialization *in/of* the cognitive processes.

In educational literature there is a variety of ways to represent knowledge—in the sense of organizing and managing it spatially.

In a desire to attempt to define the general function of spatial representations, one might affirm that they should be able to "guarantee clarity, sharing, the ability to be compared, and the capacity to evaluate the selected units of knowledge and the logical relationships outlined" (Fogarolo and Guastavigna 2013).

In this present inquiry only a few types relating to the context of education will be analyzed: two of the most widespread and oft-used in the field of didactic pedagogy and one that has recently been scientifically affirmed—though its study dates back 20 years.

M. Sorrentino (✉)
Philosophy Section, Department of Humanities, University of Naples Federico II,
via Porta di Massa 1, 80133 Naples, Italy
e-mail: sorrentino_monica@libero.it

F. Santoianni (ed.), *The Concept of Time in Early Twentieth-Century Philosophy*,
Studies in Applied Philosophy, Epistemology and Rational Ethics 24,
DOI 10.1007/978-3-319-24895-0_3

We are speaking of the model of *Mind Mapping* invented in the 1960s by the British scholar Buzan (1982; Buzan and Buzan 2000), of the theoretical operative layout of *Concept Mapping* proposed in the 1970s by the American Novak (1977, 2012) and of the *Theory of Elementary Logics* elaborated recently by the Italian Flavia Santoianni (2011, 2014).

The parallel—and sometimes intertwining—analysis of these three interpretive perspectives could serve as an epistemological scaffolding allowing us to look out on a holistic experimental paradigm on the topic of spatialization *in/of* the cognitive processes.

The graphic expressions differ on the basis of the cognitive *purpose* they pursue, in the specific logical–visual *model* they represent and in the *rules* of composition that guarantee their external and internal congruences.

In addition to the above-mentioned evaluative parameters, in the elaboration of the present interpretive parallelism there will be considered also the cognitive *role* carried out by each model, the *graphic transposition* understood as the distinctive spatial translation of a concept, the *cognitive passages* underlying visualization, the *interpretive perspective* that substantiates the model itself, and the *types of relations* that characterize it.

The cognitive scope of mind mapping is to make explicit the links and relations between concepts. Its use, in fact, can serve to activate the operative-visible dimension of mental activity and the increase in *performance*—not only those of a "linear" connotation—of conscious cognitive processes. In this sense, this means of graphic representation of knowledge would prove efficacious in summarizing others' thought by systematizing it and in creating and organizing one's personal thought—developing one's creative capacities more than one's organizational ones.

The purpose of concept mapping is the making explicit of the propositional structure of concepts so as to allow us to

> (…) represent some of the propositional structures or structures of meaning that an individual possesses for a given concept or a set of concepts (Novak 2012: 77).

Another cognitive objective at which this type of graphic expression aims is that of progressively constructing meaning through the linking of new knowledge with that already in a subject's possession in their cognitive structure when learning.

By following this logical–visual strategy, one would come to an overlapping/coincidence of the *terms* proper to a subject's compositional syntax: the purpose and the link, that is, would converge toward the definition of the trinomial "concept-relation-concept".

The cognitive purpose of the Theory of Elementary Logics instead distances itself in part from the aforementioned graphic modalities, involving, in addition to the explicit dimension of cognitive processes, also the implicit dimension.[1] It

[1]The "implicit" meaning assigned to knowledge in this interpretive frame shifts from that envisioned by scholars Nonaka and Takeuchi who in their studies (1995) consider implicit knowledge as "subjective knowledge" that must be grasped, preserved, and transformed into explicit knowledge—objective knowledge, an expression of rational thought.

therefore configures itself as an interpretive instrument useful in serving as a *trait d'union* between the explicit—understood as linguistic and mathematical thought and their relative tasks/activities—and the implicit—considered as the set of elementary operations/logics, of their relations and of their possible systematization in modifiable classes.

Therefore, if a *first* difference between these three theoretical layouts can be identified in the different purpose[2] that each pursues, a *second* distinctive aspect—closely linked to the first—can be found in the role held by each.

The declaredly "explicative" intention of mind mapping[3]—where the links "emerge" through a concrete visual outlet observable in the mental map—contributes to configure it as a co-built process of guided "making explicit", successive to the conscious phase of the analysis of conceptual relations. It in fact carries out an explicative, simplifying, dispensatory, compensative but also creative and "reinforcing" role: that is, the repetitiveness in its use would serve to "reinforce" the traceability of a cognitive pathway. In this sense some recent studies on the synaptic mechanisms of the brain would support the idea that a greater *practicability*—understood as facility in formulating a thought—of the "memory traces" (or mind mapping) would be given by a lesser biochemical/electromagnetic resistance.

In other words, one could affirm that the more one uses and repeats a schema or a map of thought, the less resistance (difficulty) will be encountered in formulating the same. This cognitive facilitating would occur precisely in virtue of the operation of reactivating (neuronal breaking through) effected along the neural pathway involved during the first journey of that specific synaptic link.

The "explicative" aim of concept mapping—from whose graphic representation there "emerges" the verbalization of the propositional architecture of concepts—translates at the operative level into a co-constructive process of (conscious) guided "making explicit" of the transformation of that cognitive structure of the knowing subject, following the phase of intentional learning. This specific theoretical–practical layout performs a "constructive" role that is not exhaustive with regard to the field of knowledge taken into consideration. In fact, it tends to furnish an instantaneous representation conscious of its own cognitive system that tendentially is continually enriched by new links with its consolidated knowledge, constructing itself in itineraries and flexibly modifying its own structure as a result of a learning process.

---

[2]Connected to the "aim", one may highlight another significant element that is useful in comparing the models considered: it is constituted by the *applicability* of the spatial representation, that is, by the hands-on application of the specific logical–visual strategy in teaching. In this sense, the use of mind mapping would prove functional to the analysis of the expressive modalities and of the representations of a text. The use of concept mapping, instead, would seem to be functional to the "revelation of the structures of knowledge possessed" by a subject (Novak 2012). The application of the theory of the *Elementary Logics* appears rather usable for the analysis of the prototypical (implicit) logical functions of one's own cognitive system, serving to structure one's personal model of study.

[3]In it the *emergence* of the links and relations between concepts—through a hub and spoke configuration—seems to take place through a spontaneous and stimulating visualization.

The "connecting" aim of the Theory of Elementary Logics—where the links between the patterns of dynamic activation, explicit and implicit, tend to remain closed within the spatial modeling, since they are not functionally necessary to their visualization—finds its application through an independent process that precedes or flanks verbalization, not following the phase of explanation through words. Its role is cognitively collaborative because it presupposes a collaboration between the two sides of knowledge, where the implicit—unlike, for example, a more exclusively metareflexive matrix that saw the role of the implicit resolved in a positive manner only if "drawn out" in the explicit—is present, precedes and is activated in an autonomous manner with respect to the explicit, which instead, if it is in difficulty, requests its help "on demand" (Santoianni 2010).

A *third* characteristic element that *distinguishes* the graphic-cognitive representation in analysis refers to their specific logical–visual configuration.

The logical–visual model of mental maps is expressed by a radiating associative visualization. It involves a laying-out of the elements of the map in a hub and spoke configuration, which reflects the "explosive" (multilateral, radiating) nature of the neurons constantly in search of new "associative" connections during the process of thought.[4]

The logical–visual model of conceptual maps is made explicit by a hierarchical visualization.[5] Its realization, in fact, presupposes the conscious choice on the part of its builder/user to identify and organize hierarchically and transversally both the concepts and their relations (conceptual learning).

The logical–visual model of elementary logics of thought translates graphically into spatializations. The latter are characterized as explicit and implicit spatial representations of the organizing criteria of concepts (Santoianni 2014). They are partially hierarchical and partially radiating: that is, the two macro-classes of *sequentiality* and *parallelism* express the highest hierarchical orders; the three classes of *union, correlation,* and *separation* denote a second hierarchical level from which branch off the six functions of *integration* (Add), *sequentiality* (Chain), *individuation* (Each), *comparison* (Compare), *inference* (Focus), and *correlation* (Link).

A *fourth* element that defines the distinctive profile of the three models of cognitive spatialization regards the rules of composition regarding the lines, the forms, the positioning, and the characters.

---

[4]Among its cognitive merits are *visibility* (concrete, spontaneous and stimulating), the *ability to be made explicit, conservation* and the *sharing* of the thought processes, the absence of the immediate search for an efficacious ordinal criteria, and the empowerment of the processes of comprehension and imagination.

[5]From the cognitive point of view one can observe how the *conceptual map* helps one to go beyond mechanical and *representative* learning—based on transmission and on the notionaistic memorization through the assigning of *labels*—precisely because it is constituted by "propositions" (understood as the main units of construction of meaning). Another positive aspect is represented by its syntax that—due to the fact of being bound, explicit, shared—favors the comprehension and evaluation of rigorous conceptual maps.

The rules of composition of the mental maps are articulated around the *Central Topic* (the initial subject matter, positioned at the center of the whole map); the *Topics* (first-level ramifications); the *Subtopics* (second-level links); the hub and spoke visualization answering to an *associative* logic; the contextual creating of a hierarchy (nonclassificatory) presupposing a *recursive* logic, the associative relations that are undefined (because they are elliptical); the labeled transversal relations; the different graphic connotations that highlight its levels of ramifications. To the rules of composition of conceptual maps, instead, there inhere the indentifying traits presented here below.

The *concept* is defined as

> regularity or schema perceived in events or objects or in testimonies/symbols/representations of events or of objects, defined by means of a label (Novak 2012: 50)

or also in terms of

> a set of meanings characterized by self-sufficiency and composed of irrenunciable elements (Fogarolo and Guastavigna 2013: 46).

Propositions understood as

> combinations of two or more words that form an affirmation regarding an event, object or idea (Novak 2012: 77).

Other compositional aspects have to do with the *main question* (which affirms its aim, perspective, and limits); the *central concept* (placed up top, at the center of the map); the *development* (that is made from top to bottom and from the general to the specific); the *inclusive* relations between concepts (indicating absolute/relative hierarchies); the *transversal* relations between concepts (connoting creativity); the *link words* (identifying/explaining the relations, traced and eventually oriented); the *link* (a concept that can be placed on the map only if it is an integral part of the load-bearing structure); the necessary *labeling* of the relations between concepts.

In addition, there appear among the compositional rules some *condition sine qua non* such as, for instance, the affirmation of the concept-relation-concept nucleus (logically complete proposition, autonomous unitary meaning); prepositions and expressions that can be placed along the relations; every concept as a starting point of new propositions; constant geometric forms,[6] types, and sizes of characters.[7]

Among the rules of composition of the *elementary logics* of thought appear cardinal concepts like that of *implicit thought* and that of *elementary logics* (Santoianni 2009, 2011).

[6]The geometrical shapes consolidate the identity and logical–visual unitary nature of the concepts.

[7]Among the other basic aspects of this model of logical–visual organization of knowledge, one may find the priority assigned to the relations and to their position on the map; the density of meaning connected to the clarity and precision of the central concepts; the typical/peculiar nature of the linking words; the validity of the things learned—in the phase of concept acquisition—and of the meanings of the propositions; the attribution of the present indicative of the verbs; the necessity of medium–long composition times (Fogarolo and Guastavigna 2013).

The first has connotations in terms of constancy, invariability,[8] persistence, prototype-like nature, generalizability, and nonreferrability.[9]

The second conceptual segment regards the existence of implicit functions of thought with a role as a prototypical connection between the explicit and the implicit elaborations. They can be expressed by spatial representations and can be configured as an instrument for linking the teaching methods, the learning processes, and the fields of knowledge.

The *implicit elementary operations* can be separated into three "classes" (union, separation, and correlation) constituted by two functions each. The "functions" are integration (Add), sequentiality (Chain), individuation (Each), comparison (Compare), inference (Focus), and correlation (Link).

More in detail, the elementary logic Add indicates *integration*: that is, the "distinctive" union (putting together) of several elements or the adding of a conceptual segment to a preexisting set of segments. It expresses the following tasks and activities: adding, subtraction, numbering, grouping, association, correspondence, explanation, and listing. The relative order of conceptual organization proceeds parallel to the integration of concepts[10] and respects the order of the learning journey.[11]

The elementary logic Chain indicates hierarchical *(con) sequentiality*: that is, the logical concatenation (cause–effect) of conceptual units. It expresses the following tasks and activities: concatenation, conclusion, implication, transformation. The relative order of conceptual organization proceeds (unlike with Add) the logical connection of concepts and allows one to control in *itinere* (unlike with Add) the phases of linking, arriving (as with Add) at a systematic and sequential linear learning. This order foresees the pre-organization of the cognitive work, the hypothesizing of logical sequences, the elaboration and the exposition that are coherent with the hypotheses.

The elementary logic Each indicates *individuation*: that is, the focusing on the conceptual element(s) present inside a set of elements. It expresses the following tasks and activities: delimiting, exclusion, distinction, analysis, division/categorization. The relative order of conceptual organization provides for the possibility of subjective analysis of the structure of the material to be known, arriving at a (systematic) analytic-distinctive learning. This order is realized through categorization/discrimination, analysis/deepening, systematization, and schematization.

---

[8]This characteristic should be understood both *horizontally*, referring in this sense to those cognitive expressions which combine various individuals, and *vertically*, linked to the idea of diachronic evolution of such expressions.

[9]This refers to the corporeal and behavioral expressions.

[10]It foresees, that is, a linear learning, gradually integrating the individual units of knowledge to the previous ones.

[11]A pathway, that is, that foresees that linguistic comprehension, re-elaboration, and oral exposition occur following the order/structure of the paragraphs and chapters of a text.

The elementary logic Compare indicates *comparison*: that is, the comparison between more than one conceptual units and the identification of affinities and differences between them. It expresses the following tasks and activities: contraposition, disjunction, comparison, quantistic comparison, and conversion of units of measurement. The relative order of conceptual organization contemplates the formation of the structures of knowledge during the carrying out of a cognitive task, developing parallel and in relation to a (qualitative and/or quantitative) comparison.

The elementary logic Focus indicates *inference*: that is, the "implicative" (inductive/deductive) origin of the concepts, one from another. It expresses the following tasks and activities: induction, deduction, interpretation, synthesis, connection, forming hypotheses, and problem solving. The relative order of conceptual organization—being linked to its possible interconnection with a multiplicity of variables—foresees that the foreman of the structures of knowledge occur in relation to the constitution of a central conceptual node, anticipating it or succeeding it.

The elementary logic Link indicates *correlation*: that is, the nonsystematic connection between units of meaning that although divergent, present possible aspects of agreement. It expresses the following tasks and activities: interrelation, linking, sharing, identifying invariants/constants, and putting into groups. The relative order of conceptual organization does not precede but follows the cognitive path: the *germinal* intuitive *idea* (Copland 1954) is followed without any logical consequentiality by links between *minor subsidiary ideas*, which are necessarily the fruit of an autonomous and personal pathway of constructing personal choices (Santoianni 2014).

A *fifth* distinctive aspect among the models (logical–visual modes of organization of knowledge) considered is represented by the process of graphic transposition of the concepts in a spatial representation.

In mind mapping and concept mapping it involves a process that is teachable by the teacher through the making explicit of all the logical–operational steps: through its use, therefore, there is realized a guided spatial construction.[12] In the case of the Theory of Elementary Logics the transpositional process is configured as nonteachable and nonconstructible (understood but not made explicit) by the one who learns, on the basis of the awareness regarding the personal prototypical logical functions. There is therefore made concrete an autonomous/self-managed spatial construction.

A *sixth* difference that emerges from the interpretive comparison refers to the cognitive passages that can be activated during spatial representation.

[12]The realization of these cognitive processes can have a *formative danger*. If the graphic representation, partially predefined, refers solely to the modalities of structuring of a specific context of knowledge and not also to the individual preferential (inasmuch as they are prototypical) modalities of cognitive functioning, not contemplating unforeseen elements of variability it can generate a situation of formative discomfort (Santoianni 2014).

In mind mapping the phases of cognitive activation begin with the activation in the user of the explicit logical functions,[13] proceeding with the reorganization of what has been spoken, and concluding with the schematization of the concepts in a graphic representation.

In concept mapping, the states of articulation of thought are that of the *formation* and the *assimilation* of an Ausubelian matrix (Novak 2012). In the context of the first cognitive process of formation there occurs the recognition of the constants of events, of objects (or of testimonies/symbols/representations of events, objects) and the use of valid linguistic labels to indicate them. During the phase of assimilation there takes place the attribution of new conceptual labels within the architecture of propositions sustained by concepts already known.

In the elementary logics the cognitive pathway is articulated around the analysis and acquisition by the user of the personal/individual implicit (prototypical) logical concepts.

A *sixth* parameter discovered by the parallel analysis of the three theoretical–practical constructs taken into consideration regards the overall interpretive perspective that each one offers.

The interpretive lens of mind mapping is of a *semantic* type, aiming at the construction of a conceptual network whose nodes are represented by "labels". The latter are keywords that express in a constant, coherent, and transparent manner the relative cognitive topics that are the objects of the specific schematization.

The epistemic perspective offered by concept mapping is of a *structural semantic* type, aiming at the progressive construction of the structural history of the meanings relative to the contents of knowledge. The compositional syntax of conceptual maps refers to the theory of learning by *assimilation* (Ausubel 1968; Novak 2012).

The central nucleus of this theory is linked to the interactive process—in the learning phase—that one observes between the materials just acquired and the preexisting *assimilative* concepts.[14]

One proceeds afterward to the "discursive" reading of the map (Novak 2012: 94) that the same author has elaborated to explain his theory.

The theory of assimilation explains human learning through the positive integration of the cognitive profile with the affective and with the psychomotorial profile. Learning can be characterized also as "mechanical" in which case it would lead to forgetting, interference, and disempowerment.

Human learning instead becomes "meaningful"—leading to empowerment—in the measure in which it respects the six *basic principles* of assimilation,

[13]With the expression "explicit logical functions" we refer to the elaborative aspect of cognitive processes that influences the intellectual styles (cognitive styles, learning styles, and styles of thought).

[14]The *assimilatory* concepts facilitate the passage of relevant information through the perceptual barriers, lightly modifying themselves and partially transforming also the information stored in the memory.

superordinate learning, progressive differentiation, integrative reconciliation, assimilation by cancelation, and anticipatory organizers (Novak 2012, 91–134).[15]

Meaningful learning constructs the cognitive structure—stored in long-term memory—which is enriched through the activation of metacognitive instruments (such as conceptual maps). Such instruments can favor the dynamics of problem solving and the various cognitive operations regarding the developmental age of the individual, favoring short-term memory over long term.

The cognitive structure is built by the structure of knowledge. The production of knowledge is described by human constructivism—which is focused on events—and is shown in the V Diagram entitled *Diagram of Knowledge* (Novak 2012: 135–182).[16]

Meaningful learning in this sense expresses the capacities to acquire, utilize, and create knowledge.

The interpretive key of the elementary logics is of a *prototypical management/operational* type in the sense that through the identification of prototypical models of thought—expressing implicit cognitive functions—it is possible to construct and manage explicit knowledge. This interpretive approach starts from the *adaptive* theory of learning through the activation of cognitive *educability* (Santoianni 2006).

The core of this theory is linked to the process of reciprocally adaptive interaction—understood as structural coupling (Riegler 2002)—that occurs in the learning phase between the individual and his learning environment.

Learning takes on an *adaptive* connotation since it considers a functional system in its *entirety*—including the aspects of emotivity, cognitive elaboration, perception, superior elaboration, the implicit and the explicit—and cognition itself in its "distributed", "situated", and "incarnate" sense.

The analysis of interpretive constructs of Novak and Santoianni has caused some converging and other divergent aspects to emerge which seem to proceed in parallel fashion toward a common horizon.

---

[15]There exist some similarities and differences among Piaget's ideas on assimilation, accommodation, re-equilibrium and Ususbel's ideas on assimilation, progressive differentiation, and integrative reconciliation in the version of it that Novak offers. The stadial nature of the cognitive development of Piaget regards a general reasoning capacity; Novak's idea that starts from Ausubel's theory is supported by the conviction that the reasoning capacities are essentially linked to the *adequacy* of the conceptual structures of a subject in relation to a specific context of knowledge. There could be linked to these perspectives (in particular, to Novak's interpretive proposal) the bioeducative approach of Santoianni that in regard to this sustains that the functionally efficacious reasoning capacities follow a logic of cognitive closeness/distance correspondence between the specific cognitive *prism* of a subject who learns and the internal cognitive structure of the relative disciplinary field he or she is dealing with (Santoianni 2006, 2010).

[16]The use of the diagram is justified by the close link that according to Novak exists between meaningful learning and the construction of knowledge. In the author's thought, in fact, the construction of knowledge represents an "extension of the human capacity to create new meanings (…)" (Novak 2012: 154).

Here below we offer the comparative elements examined.

Novak's theory of meaningful learning regards the progressive construction of meaning through the linking, grafting, modification, and transformation of new knowledge and that already "assimilated."

Santoianni's theory on adaptive learning regards the construction of the personal cognitive system (*cognitive prism*), an expression of the adaptive relation between the explicit dimension (educability) and the implicit dimension (prototypical nature, spatiality, generalizability) of knowing.

Meaningful learning involves the individual (at the cognitive, affective and psychomotorial level), the context, the concrete experience (objects/events), and knowledge (explicit: fruit of the active (constructivist) interaction between *thought* —conceptual aspect—and *action*—methodological aspect).

Adaptive learning involves the individual (at the cognitive, metareflexive, intersubjective-relational, affective-emotional, organismic, implicit level), the adaptive system, and knowledge (explicit, implicit).

The comparison between these two theoretical–methodological layouts would seem to highlight some shared premises on the development of the cognitive structures.

In this sense, from such an epistemic-comparative examination, it would emerge that learning and development in a child, though in direct relation, are never realized in the same way or parallel to one another. One could thus affirm that between learning and processes of development there exist complex dynamic relations that cannot be categorized in terms of "unmodifiable theoretical formulations" (Vygotskij 1986: 91; Novak 2012; Santoianni 2014). Another aspect to emphasize regards the differences that can be found between the cognitive systems of children and those of adults that would not seem to effectively and radically reflect the various *stages* of Piagetan development. In this shared horizon of research therefore, both the minds of children and those of adults would vary qualitatively in terms of cognitive performance in relation to contexts and circumstances (contextually specific competencies). The cognitive implication that one deduces is that over the course of infancy and adolescence, the linear and concise identification of "stages" of development of thought (Piaget 1926) would no longer seem sustainable, in virtue of the documented affirmation of the existence of *tendencies* of development (Flavell 1985: 114; Novak 2012; Santoianni 2006, 2010, 2014).

An *eighth* distinctive aspect refers to the type of relations that the considered models of spatial representation express.

Mind mapping expresses derivative relations of a deductive/inductive kind, concept mapping expresses causal relations with a sequential/comparative connotation, the elementary logics express integrative, sequential, identifying, comparative, derivative, correlative relations.

References

Ausubel, D. (1968). *Educational psychology. A cognitive view*. New York: Holt, Rinehart & Winston.
Buzan, T. (1982). *Usiamo la testa*. Milano: Frassinelli.
Buzan, B, & Buzan, T. (2000). *The mind map book*. London: BBC Books.
Copland, A. (1954). *Come ascoltare la musica*. Milano: Garzanti.
Flavell, J. (1985). *Cognitive development*. Englewood Cliffs: Prentice-Hall.
Fogarolo, F, & Guastavigna, M. (2013). *Insegnare e imparare con le mappe. Strategie logico-visive per l'organizzazione delle conoscenze*. Trento: Erickson.
Novak, J. (1977). *A theory of education*. Ithaca: Cornell University Press.
Novak, J. (2012). *Costruire mappe concettuali. Strategie e metodi per utilizzarle nella didattica*. Trento: Erickson.
Nonaka, I, & Takeuchi, H. (1995). *The knowledge-creating company*. New York: Oxford University.
Piaget, J. (1926). *The language and thought of the child*. New York: Harcourt Brace & Company.
Riegler, A. (2002). When is a cognitive system embodied? *Cognitive Systems Research, 3*, 339–348.
Santoianni, F. (2006). *Apprendere al singolare, insegnare al plurale*. Roma: Carocci.
Santoianni, F. (2009). *Costruzione di ambienti per lo sviluppo e l'apprendimento. Il protocollo formativo C.A.S.A. per la scuola primaria*. Naples: Consorzio Editoriale Fridericiana.
Santoianni, F. (2010). *Modelli e strumenti di insegnamento: approcci per migliorare l'esperienza didattica*. Roma: Carocci.
Santoianni, F. (2011). Educational models of knowledge prototypes development. *Mind & Society, 10*, 103–129.
Santoianni, F. (2014). *Modelli di studio. Apprendere con la teoria delle logiche elementari*. Trento: Erickson.
Vygotskij, L. (1986). *Thought and language*. Cambridge: The MIT Press.

Part I
Phenomenology and Perception of Time

Introduction

Alessandro Arienzo

Abstract From Aristotle to Gilles Deleuze and Felix Guattari time has been interpreted in a number of ways. The chapters presented in this section mostly discuss contemporary philosophers who were influenced by Bergson and that may be broadly inscribed within the phenomenological approach. In the introduction we will focus on a few key readings of time, focus in on Aristotle, Augustin, Henry Bergson and Reinhardt Koselleck in order to point out how the nineteenth and twentieth centuries would develop further the notion of time by multiplying not only its definitions and descriptions but time itself.

"Time is number of movement in respect of the before and after": this is the most quoted passage of Aristotle's extended discussion of time that occurs in *Physics* (217 b 30–224 a 17). Time is a determination of motion and its study is part of a broader investigation on physics and corruptible bodies.

Aristotle's discussion of time focuses on three key problems: the existence of time, the nature of the present (now), and the nature of time itself. According to the Stagirite, to grasp the essence of time it is necessary to understand the relation between the time present, in which I experience time, and change, which can only be experienced by the motion of things. In fact, it is only the changes in nature and in the order of things that make men able to experience the continuous sequence of "nows" as a succession of events from time-no-longer to time-not-yet. In this respect time has a double nature, it expresses movement and change, but it also represents and gives measure to the natural alteration of things and their movements for "every alteration and all that changes is in time" (222 b 31). In the book of his *Physics* dedicated to space, number and motion, time appears as an ordering principle: it gives order to space through movement. As change is experienced in the passing of nows, change is in its own way motion.

The intertwining of time and change is expressed by Aristotle with great clarity: "not only do we measure change by time, but time by change, because they are

A. Arienzo (✉)
Philosophy Section, Department of Humanities, University of Naples Federico II,
via Porta di Massa 1, 80133 Naples, Italy
e-mail: alessandro.arienzo@unina.it

F. Santoianni (ed.), *The Concept of Time in Early Twentieth-Century Philosophy*,
Studies in Applied Philosophy, Epistemology and Rational Ethics 24,
DOI 10.1007/978-3-319-24895-0_4

defined by one another" (220 b 14–15). However, time is not identical to change and movement for those are always "in space" while time has no spatiality, moreover changes and movements involve differences in speed and velocity that do not belong to time. Certainly, in order to measure time it is necessary to have a unit of measure and this unit can only be found, according to Aristotle, in the uniform, perfect, and circular motion of stars and celestial bodies.

A different approach to the study of time was introduced by Augustine of Hippo, who distinguished between the immote time of God and human time as the distending of the soul (*distentio animae*). He deals with time in Book XI of the *Confessions* in which he explains that God is eternal—he does not take action at individual moments in time, because he does not exist in individual moments in time—and is outside of time. This leads to the question "what exactly is time?" and to his famous affirmation that "If no one asks me, I know what it is" (Book XI chap. XIV 17). Certainly, time has three dimensions: the past, the present, and the future, and among them the past and future do not actually exist, since we can only experience them in the present. Augustine is not questioning the basic idea that time itself exists, since it can be measured, but the measuring of time leave open the issue of the nature of past and future. He only observes that the present must become past in order to be time (otherwise, it would be eternity) and that time present can only come into existence when it passes into time past. Similar questions thus arise with time future. Augustine was therefore convinced that time does not exist in the physical world and that it only exists inside the mind, for "It appears to me that time is nothing other than extendedness; but extendedness of what I do not know. This is a marvel to me. The extendedness may be of the mind itself" (Book XI chap. XIV 33).

Modern philosophical historiography has often interpreted Augustine in contrast to Aristotle. The Stagirite was interpreted as the advocate of an objectivist theory of time while Augustine was described as the supporter of a subjectivist theory. Without disputing the differences between the two, we can also find in the Aristotelian analysis of time an important relation between the three dimensions of time (earlier, now, and after) and the "soul" of men. In fact, according to Aristotle, the perception of before and after, thus of number in motion, necessarily presupposes the soul: "When we think of the extremes as different from the middle and the mind pronounces that the 'nows' are two, one before and one after, it is then that we say that there is time, and this that we say is time. For what is bounded by the 'now' is thought to be time—we may assume this" (*Phys.* D 11). In other words, the soul is the spiritual counting principle and the precondition itself of a distinction between things measured and the measure, thus, of time: "Whether if soul did not exist time would exist or not, is a question that may be asked fairly; for if there cannot be some one to count there cannot be anything that can be counted, so that evidently there cannot be number, for number is either what has been, or what can be, counted. But if nothing but soul, or in soul reason, is qualified to count, there would be no time unless there were soul, but only that of which time is an attribute, i.e. if movement can exist without soul, and the before and after are attributes of movement, and time is these qua numerable" (*Phys.* D 14).

Medieval and early modern philosophy, by following or contrasting Aristotelianism and/or Augustinianism, would enrich the ways in which time was interpreted. Indeed, for a long time common people would continue to experience time as circular, representing the endless natural motion of beings, flowing up to the eschatological end of earthly corruption with the call of the world to the infinite presence of God. Time was thus experienced as the natural flow of things in its contrast with the stable eternity of God. In this respect, it joined the external features of movement and measure and the internal motion of the soul. After having characterized ancient, medieval, and modern history of western philosophy and science, Aristotle's and Augustine analysis are still today at the core of contemporary investigations into time, both in philosophical and scientific enquiry. Certainly, a number of relevant advancements occurred in the philosophical investigation of time. In the early modern age the affirmation of quantitative and experimental sciences, the rupture of the Christian religious unity, the discoveries of the new world, the development of mechanics and manufacturing (clearly represented by the improvements in the production of clocks) are among the many reasons time started to acquire not only a linear and quantitative description but also a progressive overtone. Representative of these changes is the meaning of the word "revolution" that from the circular motion of orbs came to describe a sudden rupture in a political order.

Reinhardt Koselleck has interpreted this massive change as a change in temporalization, which is to say that the experience of a specifically "modern" time is, in reality, the experience of a plurality of "times". He also reminds us that the circular, natural conception of time was slowly replaced by a time in which human reason was deemed to progressively perfect itself. Gottfried Wilhelm Leibniz, among others, has put forward the thesis that the universe neither repeats itself nor ages, nor will it even ever reach any point of completion, of maturity. In this sense, he formulated a dynamic conception of time in which temporality was nonetheless inherent in progress. Moreover, a new sensibility toward history characterized western culture from the Renaissance to nineteenth century that Koselleck described as a change in the "space of experience" and in the "horizon of expectation". With the former expression, Koselleck describes the complex of human experiences shaping the past and present, while by the latter he means human pretention toward the future through expectations, plans, desires. In western modern, culture, what he calls the "space of experience," has become ever more distant from the "horizon of expectation." Experience has become less and less relevant to foretelling the future for it is deemed impossible to extend one's knowledge of the past into an expectation of what is to come. Historical developments on the scale of the French and Industrial Revolutions destroyed what had been an age-old confidence that the future could be known on the basis of the past. The acceleration of time and its opening up to an uncertain but progressive future is therefore the main feature of modernity.

The nineteenth and twentieth centuries would develop further the notion of time by multiplying not only its definitions and descriptions but time itself. Logic, physics, and psychology are probably the sciences that most fostered this change and that greatly influenced philosophical reflections on time. The third section of this volume deals with the decisive changes in the study of time that were stimulated by scientific research in the early twentieth century, an influence that is well exemplified by Einstein's theory of relativity. As far as philosophy is concerned, among the most influential thinkers, at the end of nineteenth century Henry Bergson distinguished between time experienced (real duration, i.e., *durée réelle*) and the time of science. The latter is based on the imposition of spatial concepts onto time, leading to a perception of time as a succession of separate, discrete, spatial constructs. His major work, *Matter and Memory* (1896), was an essay on the relation between mind and body wherein he argues that it is only in the memory that the reality of mind and the reality of matter converge. According to Bergson, the linear duration of time is a fiction for there is no single temporality but many different rhythms in time which, slower or faster, relax or thicken our consciousness of time. While space is somehow homogeneous, the duration and succession of time are not, as they belong to the conscious mind.

Bergson will deeply influence the many different "continental" philosophical approach to time, from the phenomenological to the existential, from Heidegger to Binswanger, from Derrida to Foucault, from Jaspers to Sartre; authors and approaches to which this chapter is mostly dedicated. A second most relevant approach to the study of time would be that of John McTaggart, whose most influential article, published in 1908, *The Unreality of Time*, is still at the core of the debates in contemporary analytical philosophy on time and its nature. In his article, McTaggart distinguished two different approaches to time. The first draws an image of time as a temporal line moving from the past toward the present to the future. McTaggart named this approach the A theory. In this theoretical framework, temporal properties are not permanent as time is expressed by a continuous condition of motion of instants. A different approach is based on a fundamental difference between "being earlier than something" and "being later than something" and he named it the B theory. In this case temporal properties are stable as they can be summed up in the two notions of "earlier than" and "later than". McTaggart's aim was to show the unreality of time while describing the paradoxes involved in those two conceptions of time. Actual debate on the nature of time stemming from McTaggart's position is characterized by three positions that can be named presentism, eternalism, and growing-past theory. According to presentists time is always a present experience even when we experience memories or future. Growing-past theories attribute reality only to past and present while future, in its indeterminacy, has only a potential nature. Instead, eternalists do not attribute any specific difference to present, past, and future time as the only difference in their experience is in the subject.

The chapters presented in this section mostly discuss contemporary philosophers who were influenced by Bergson and that may be broadly inscribed within the phenomenological approach.

Mirko Di Bernardo discusses the problem of time in the thought of Henri Bergson, focusing on his analysis of duration as it is conducive to a philosophy of intuition.

Felice Masi focuses on the concept of temporality as it is discussed by Edmund Husserl as the key concept of its phenomenology of time. In his paper, Masi also present the most relevant critiques to Husserl made in the twentieth century by philosophers such as Heidegger, Derrida, Bergmann, and Lèvinas.

Armando Mascolo reads Jean-Paul Sartre in his relation to Heidegger by focusing on the structures of consciousness interpreted as the conditions of man's absolute freedom.

In her essay on Merleay Ponty, *Maria Teresa Catena* discusses the nexus between time and its perception focusing on the relevance of the body as the forefront of subjectivity.

Alessandro Arienzo discusses the critique made by Gilles Deleuze and Felix Guattari to the linear conception of time developed in traditional Marxism and offers the basic tenets of their conception of time in an effort to understand and oppose capitalistic society.

Lastly, Derrida's deconstruction of western metaphysics of presence is the subject of *Marco Stimolo*'s paper. In order to investigate the implication of Derrida's deconstruction of the transcendental signifier for scientific inquiry, the example of the economic analysis of individual behavior is discussed.

Phenomenology and Perception of Time Maps

Flavia Santoianni

1 Aristotle and Augustine of Hippo Atlas EL MAP

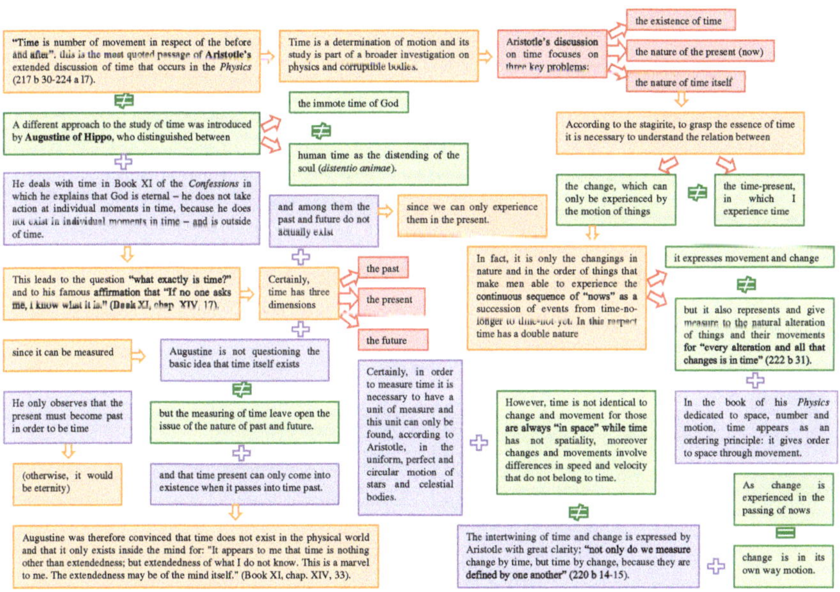

Time in Aristotle and Augustine of Hippo – EL map – Flavia Santoianni

F. Santoianni (✉)
Philosophy Section, Department of Humanities, University of Naples Federico II, via Porta di Massa 1, 80133 Naples, Italy
e-mail: bes@unina.it

F. Santoianni (ed.), *The Concept of Time in Early Twentieth-Century Philosophy*, Studies in Applied Philosophy, Epistemology and Rational Ethics 24,
DOI 10.1007/978-3-319-24895-0_5

2 Bergson, Husserl, Sartre, and Merleau-Ponty Atlas EL MAP

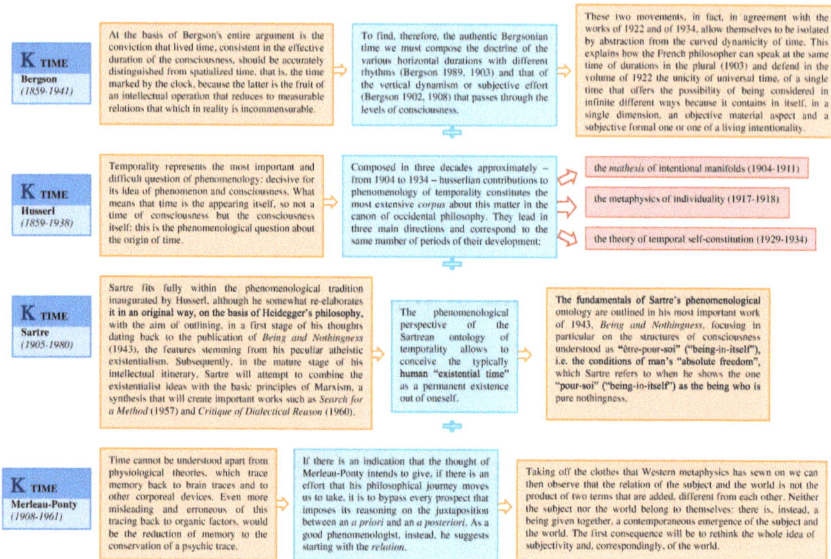

Time in Bergson, Husserl, Sartre, and Merleau-Ponty – EL map – Flavia Santoianni

3 Deleuze, Guattari, and Derrida Atlas EL MAP

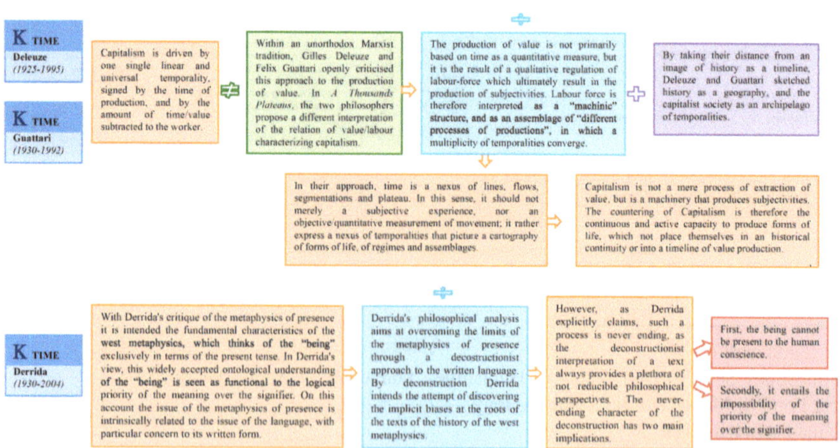

Time in Deleuze, Guattari, and Derrida – EL map –Flavia Santoianni

4 Koselleck and Mctaggard Atlas EL MAP

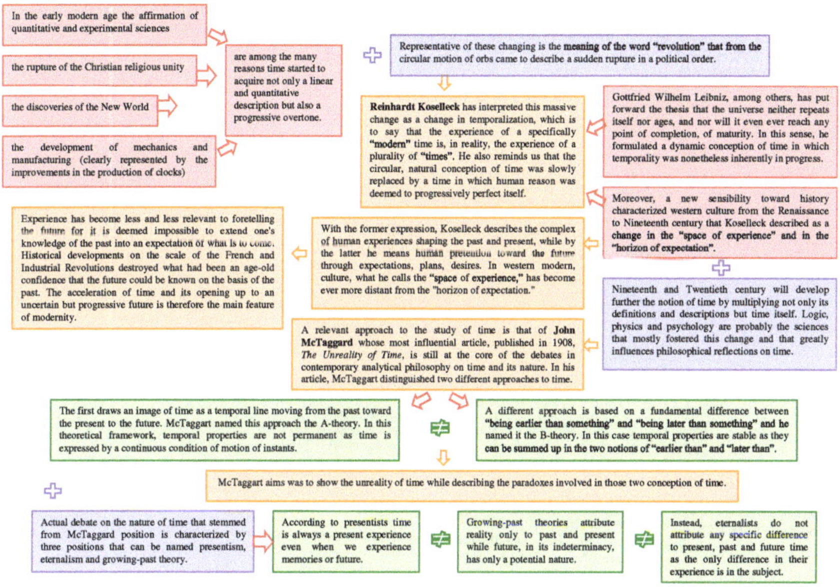

Time in Koselleck and McTaggard – EL map – Flavia Santoianni

5 Atlas Map

Atlas map –*Philosophers are located near their birthplaces. The Geographical Boundaries of Countries May Differ in Comparison with the European Geography of the Early Twentieth-Century*

Time and Reality in the Thought of Henri Bergson

Mirko Di Bernardo

Abstract This chapter discusses the problem of time in the thought of Bergson, showing how the evolution of the concept of duration is conducive to new developments in the philosophy of intuition. Duration, which in the *Essay* connotes the experience of a non-measurable lived experience, while in *Matter and Memory* it assumes rhythms of different intensities to justify the relationship between perception and memory, as well as in *Creative Evolution* is judged as the fabric of reality itself, in *Duration and Simultaneity* it is posited not only as a criterion to discern what is real and what is artificial, but also to justify the measurement of reality, that is to say, to restore the point of contact between time as duration and space, that seemed previously compromised by the loss of ontological consistency of extension as the dimension of the body. In this paper, then, Bergson, also by virtue of a controversial, detailed comparison with the theory of relativity, finally arrives to support the hypothesis of a single temporal dimension, curved and dynamic, where space comes to be outlined as the abstract tangent of time. It is at this level, then, that it becomes possible to examine the emergence of the concepts within the meanders of intuition as concepts imbued with meaning: here is the primary source of that continuous "added" element of new nuclei of creativity that characterizes the very logic of Bergsonian living where what, as ideal relationship, is time, as a real relationship, becomes life, i.e., the continuous opening of a register in which time is inscribed, becoming incarnate thus in the very nature of every organism.

M. Di Bernardo (✉)
Department of Business Government Philosophy Studies, University of Rome "Tor Vergata", via Columbia 1, 00199 Rome, Italy
e-mail: mirko.di.bernardo@uniroma2.it

F. Santoianni (ed.), *The Concept of Time in Early Twentieth-Century Philosophy*, Studies in Applied Philosophy, Epistemology and Rational Ethics 24,
DOI 10.1007/978-3-319-24895-0_6

1 Real Duration

The central thrust of Bergsonian philosophy is constituted by the doctrine of the real (or concrete) duration formulated in the volume of 1889 with the title: *Essai sur les données immédiates de la conscience*, drawing inspiration from the evolutionary philosophy of H. Spencer. We may say that for Bergson real duration is presented as the fact of consciousness stripped of every intellectual superstructure and recognized in its originary simplicity. The psychological instrospection, restored to its most authentic philosophical meaning and liberated from every deforming intellectualism is here taken up as the basis of a metaphysical vision of universal reality. With this, however, the French philosopher does not understand duration as something that always remains the same. Quite the contrary: "to last," for Bergson, means "to exist as consciousness": the consciousness "lasts," remains, continues to be precisely because it changes, shifts and transforms continuously. From this, he matures the necessity to give an adequate definition of time, one capable of overcoming the Spencerian solution, according to which the nature of time is in itself unknowable, as well as overcoming the scientific solution, according to which time is an indefinite succession of homogeneous and uniform instants.

In contrast with the scientific psychology that presumed the reduction of the I to a succession of psychic states linked according to determinate laws and relations, for Bergson the reality of consciousness, if one tries to grasp it in its most profound essence or interior life and not in its exterior manifestations, is revealed instead as an incessant current, of a purely qualitative nature, whose moments are not reciprocally juxtaposed, but blend and co-penetrate in an organic and incompressible whole similar to the flowing of a melody in which each note is prolonged in the others. This is real duration, that is to say, a reality in continuous renewal that, resolving in itself its passing and creating that which is to come, is deployed as a living process, "uninterrupted spring of newness" (Bergson 1889), from which life flows incessantly. At this level, therefore, time comes to be conceived as a succession of qualitative states of consciousness, all different but intimately connected.

The idea of duration is presented in an articulate manner in the second chapter of the *Essai* where Bergson contrasts the multiplicity of material objects, a multiplicity that takes on a numerical form and is arranged ideally as a juxtaposition, that is as a whole of elements set side-by-side in linear fashion (just like a numerical series), with a confused multiplicity of immaterial and successive elements, proper to the states of consciousness. The authentic temporality of consciousness that is designated here by the French philosopher with the term "duration" cannot therefore be expressed by that concept of time, familiar to common sense, that has found its technical re-elaboration in physics. The time of which science speaks, in fact, is a construction of the intelligence and as such carries with it a spatial dimension. Not by chance does science represent time through recourse to geometrical figures such as the point and the line, preferring static symbols to measure that which changes (Pessina 1994). Such a time in reality is nothing other than a ghost of space since it is born from the introduction of spatial representations into the evaluation of the facts of consciousness, which has no spatiality because the pure duration in which the concrete life of the I consists is a reciprocal co-penetration and solidarity in which each moment is not outside the other, but in the other. What is affirmed about time, in agreement with Bergson, can also be said about movement, where one must not confuse the space travelled, which is a homogeneous (measurable) quantity with the act that travels through it, which is pure (indivisible) intensity. This ancient confusion between time and space generated, according to Bergson, the aporias of Zeno, which in the second half of the nineteenth century were revisited precisely from a scientific point of view. Time can be thought as homogeneous only if it is conceived of through space which is uniform. Interior duration, authentic time that implies memory as awareness of the effective progress of the life of the I, is instead a qualitative multiplicity that is not deployed in space, but is concentrated in the very act of changing. Therefore, unlike spatialized temporality represented by the measuring of the watch whose successive oscillations exist only for a conscious external observer who remembers the past and juxtaposes the symbols of the two oscillations in an auxiliary space, the time lived as duration consists in "a process of organization or of mutual co-penetration of the facts of consciousness" that allows the formation of representations of the oscillations that have passed in the same time in which the current oscillation is perceived (Bergson 1889: 71–72).

The main error of scientific psychology, therefore, consists in the pretension of subjecting to measurement the psychic facts as if they were an object of space; in the pretention, that is to say, of introducing into the evaluation of the facts of consciousness concepts that can have an explicative value only for the phenomenal world, thus creating a "mathematical time" considered as a homogeneous *quid* in which psychic facts, cut off from their living concreteness, are fixed and almost solidified. In Bergson's eyes, instead, the multiplicity of psychic states, that psychology addresses, has nothing to do with the numerical multiplicity or with the multiplicity of physical entities: the life of consciousness, in fact, proves unspeakable with the categories of the intellect. In the psychic life that which is usually called multiplicity corresponds in reality to the qualitative variations, to the co-penetration of sensations, sentiments and thoughts that constitute the I (Pessina 1987, 1994). Duration, therefore, understood as lived human experience, can manifest itself not through an intellective act (the extensional dimension of the spatialization of time), but in virtue of an intuition (the hyper-intensional and intentional dimension of the temporalization of space). This is a thesis that will only later find more precise gnoseological developments (Bergson 1903, 1907, 1934) but which already starting from the *Essai* indicates the condition for overcoming the intellectualistic level of knowledge (Pessina 1987). These themes induce Bergson to distinguish clearly the examples of the predicting of the future furnished by science from the real development of the conscious life. The scientist, in fact, limits himself to establishing the relationship between given intervals, but real duration escapes his calculations. In other words, one can express the time elapsed in spatial terms, but not the time that elapses, that in which the free (intentional) act is produced. Freedom, like consciousness itself in its living actuality, is therefore, according to Bergson, an undeniable reality, but in vain do we seek to give it a definition without de-naturing it, translating it into an unintelligible language in which the codes constantly flee from every attempt at a complete determination of them. The duration of consciousness, in fact, is an infinitary process in continuous evolution, it is a parade of shadings capable of keeping together continuity and heterogeneity where the former has an onto-semantic priority over the latter. Duration as integral conservation, constant development that, deploying itself in virtue of the contribution of a memory that conserves everything, reveals itself as the generation of continuous novelty, as the expression of a temporal depth that in its emergence in the surface becomes flesh, organic life that develops temporally and in which memory, the metaphysical heart of duration that marks the path, generates heterogeneity and meaning.

2 The Doctrine of the Multiplicity of Levels

In 1896 *Matière et mémoire* saw the light. The work was dedicated to the study of the relationships between the body and the spirit, whose essence is located in the *memory*, while the body is attributed with the function of choosing and limiting the memories for the purposes of action. It is an outlook that would have profound consequences also with regard to the problem we are addressing: consciousness, in fact, is seen as freedom, its interior time is unforeseeable, the time of the "profound I," opposed to the "superficial I" subjected to the automatisms of praxis and of conventions. Therefore, without memory there would not even be the intuition of duration, that is, of the flowing of psychic life (Taroni 1999: 79–85). One could, then, say that memory is consciousness and is, at the same time, duration. However, unlike the *Essai* in which the stratification of being while being required by the coherence of the whole does not appear except in sporadic glimpses, in *Matière et mémoir* it dominates the scene, finally being explicitly enunciated in the final chapter where Bergson uses it to systematize within the harmonious whole the results of the work together with the doctrines elaborated in the previous works. In brief, the conclusion is that duration occurs in "different rhythms" that measure the level of tension or relaxation of consciousness, thus fixing "their respective places in the series of beings": in fact, in stretching out it comes close to the mode of being of material things and in concentrating itself is likened to those of the spirit and the continual "infinity of levels" that develop between the two extreme modes of being constitutes the real "relation between the body and the spirit" that the subtitle of the volume proposes to investigate (Bergson 1896: 190–208).

In reality, the doctrine of the multiplicity of levels referred to the multiple moments of duration is based on the idea according to which memory is not only conservation (an operation that allows to leave unchanged the successive visions while not impeding one from distinguishing one from the other), but contracts and concentrates the scattered elements of experience—an operation of fusion of successive moments of duration that in becoming a single state of consciousness undergo a profound qualitative transformation.[1] There is, therefore, a stratified duration that at a given level is slower and, at the same time is tenser. One thus has multiple durations that constitute metaphysically different levels of being: sensation, for example, is more contracted than spatial exteriority and sentiment more than sensation. In this guise, as one descends deep down, one verifies an increase in concreteness and a loss of objectivity, finally reaching the extreme limit where one has, according to Bergson, the "eternity of life" in which every time is as if gathered up into an absolutely concentrated point (a principle that "takes together" all the elements of experience) where there are preserved the memories that can become present in any moment: the human being, in fact, can remember only in the measure in which he has interiorized (conservation of the past by contraction) experience deep down, an experience, that is to say, that through an articulated and stratified dialectical process with a dual internal selection, in turn, will stretch out from deep down as a memory through the level of sentiment—the nostalgic dimension.[2] The distinction of the levels of duration thus offers a new structure for the doctrine of the conservation of the past in the present already affirmed by the *Essai*. If there were only a single level of duration, the past would be resolved entirely in the present in the sense that it could flow into it, bringing there the material preserved, but it could not be present to it as past. However, in *Matière et mémoire* Bergson shows a more complex relationship because he highlights the fact that the past is not resolved in the present since it coexists with it on another level of duration, a deeper level and one of a tenser rhythm.

Given this, therefore, following the phenomenological examination of the gradually deeper layers of the human being that lead Bergson to formulate a hypothesis on the manner with which the past is conserved, it appears clear how the

[1]In the *Essai*, a single apperception was supposed, able to embrace a long series of events, leaving them however as they were without transforming them qualitatively. The contraction of which *Matière et mémoire* speaks instead is very different from a sum, because it does not leave the phenomenon as it is, but transforms it into an absolutely heterogeneous quality. Here the contracting memory does not have the function of preserving, but that of leading to "different and irreducible levels of duration" (Mathieu 1954).

[2]The punctual and concentrated preservation of memories does not appear to be in contradiction with the image of the inverted cone whose vertex is in the present and whose base is formed by memory, used by Bergson himself in other passages of *Matière et mémoire*. The cone, in fact, represents only the diminishing of the possible field of consciousness with the concentrating of attention and is not in contrast with the idea of memories that persist, all gathered up into a point.

"eternity of life" (infinitely contracted duration-limit) and space[3] (diluted and dispersed duration-limit) are the two extreme principles of human experience between which it proves possible to find a continual gradation of modes of being (Bergson 1903). No precise cut, therefore, between "objective primary qualities" and "subjective secondary qualities," but a continual passage that transforms this distinction. In other words, it involves a passage in which reciprocal spatiality and exteriority diminish while the subjective character increases, that is, the level of tension of duration with which the elements of experience are taken up in the subject. The Bergsonian interior dynamism opens then to a new dimension in which the duration can no longer be ascribed to the linear succession or to the spirit in its progressive actualization (Bergson 1896), but comes to be configured as a continual range of durations of different rhythms but united by a movement that is not identified with any of them since it develops in a different direction, in a direction, that is, perpendicular to them so as to connect together the various durations that are little by little developing in parallel fashion (Bergson 1896, 1902, 1903). This movement, in Bergson's eyes, coincides with the spirit of every human being. Every person, in fact, finds his own deepest root in an infinitely contracted duration that transcends his experience while, on the physical level, it adheres with its own surface to an infinitely extended duration (extended exteriority): the intermedial levels between these two extremes are kept together by the activity of the human spirit that, passing continually from the depth to the surface and vice versa, passes through the different levels of being contributing, in a certain sense, also to their constitution. The human being, according to Bergson, is essentially this act, an act, that is, which gives concreteness to the different durations which per se would be nothing but pure abstractions (sections cut ideally in a unitary whole). Thus with respect to the movement that is carried out between the before and after at a determinate level of duration (a horizontal dynamism since it is carried out at a constant distance from a center), the movement of the spirit between the depth and the surface is defined by Bergson as "vertical dynamism" as it is directed towards a center or moves away from it. With regard to the *Essai*, where the horizontal perspective is mainly developed, in *Matière et mémoire* and above all in the collection of essays and conferences subsequent to it (but gravitating around the same questions), gathered in the volume from 1919 *Energie spirituelle*, the French philosopher attempts to fix some characteristics of the vertical movements (from the depth towards the surface)

[3]If one does not want to return to the immanence of the *Essai*, where space was an absolute originary principle that permitted denying existence to spatial things, which the doctrine of the multiplicity of levels of duration has overcome, it is necessary however to admit, according to Bergson, as the rhythm of duration gradually slows, an ever greater reciprocal exteriority and homogeneity between the moments of duration, that is, a tending towards the limit of spatiality. Spatiality, therefore, in *Matière et mémoire* exists as a limit and is the principle by which the moments of the various durations can be distinguished, though never absolutely, from each other. Thus, though without pretending to glimpse in space a concrete object of experience, but only a principle, the work from 1896 not only does not forbid, but even imposes, in order to avoid falling back into absolute immanence, presupposing this principle so that the considerations developed on space in 1889 continue to be valid.

finally coming to delineate the theory of vertical dynamism already mentioned in an explicit manner in the essay of 1902 with the title: *L'effort intellectuel*, where the attention is focused not on the intensity, but on the intrinsic characteristics of the spiritual acts accompanied by a sense of fatigue (for example, the effort to remember, understand and invent), that is, on their nature. The intellectual effort is a movement, which Bergson calls vertical with respect to the horizontal levels of duration, with a "dynamic schema" in the direction of the image that develops it, it is a continual transformation of abstract relations suggested by the objects perceived in concrete images capable of covering such objects: it involves, in other words, traversing different rhythms and levels between the depth and the surface while keeping distinct the difference between temporal relations between the various mathematically representable movements (quantitative schema or skeleton of the image) and the qualitative impression (dynamic schema or the interiorized image itself) which cannot be formalized because it is pure incompressibility. The coincidence of the quantitative schema and the dynamic schema, in fact, would annul precisely that space that Bergson assigns to the intellectual effort (movement towards the image) so that it develops, permitting, moreover, the image to surface in the mind. It is clear then how the spiritual acts, in traversing different levels of duration, cannot be ascribed to the psychological processes that come close in some way to mechanical phenomena, being carried out on the same level of consciousness without involving the depth dimension. The real spiritual acts instead are those psychological phenomena that are deployed in a vertical movement and in which there is a creation of novelty or at least an effort (Bergson 1902).

Thus, if the most profound way of being was represented by the first Bergson with a point, that is, with the absolute spatial negation of space ("via negativa"), now it is expressed positively through the identification with a pure dynamic thrust: the most profound point of our being is a "pure going towards" not yet widened in its object (Bergson 1902, 1903, 1908). Therefore, what we grasp in the deepest layers of our consciousness is rather a dynamic thrust that, though not always pure, draws so close to the origin as to allow us, by passing the limit, to indentify in any case, though only in phenomenological terms, the most profound mode of being: it is an indicating, a tending, an "intention." It is the meaning of something without yet having the thing signified. In other words, it involves a very precise meaning (we are able, in fact, to judge immediately whether a given exterior form is suitable to it or not) which however, as a very determinate impression as "tending towards," does not allow itself to be defined as an object in itself, eluding every attempt at grasping it (Bergson 1902). According to Bergson, therefore, every "I", in the deepest point, is a "tending towards" gradually less profound levels of duration and, ultimately, towards the level of objectivity (the opposite level). In this sense, with respect to the various levels on which the duration develops, this movement begins with perpendicular, or instantaneous being. It involves that typically instantaneous beginning of the spiritual act that in *Le souvenir du présent et la fausse reconnaissance* assumes the transcendental characteristic of memory that exits time, all levels of time, after having traversed them[4] (Bergson 1908) and that sometimes more generally has been represented with the image of the brainwave. This dynamic movement, however, gradually expanding in the dimension of objectivity, develops (uniting the most superficial layers of being to the most profound) in an extended duration where the delay that the effort implies regards the deployment of the spiritual act itself (Bergson 1902).

Deep down we have, therefore, according to the French philosopher, a nothingness of objectivity and a maximum of activity directed towards the object: an intention that is impossible to grasp in its originary purity, but revealed by the way in which the functions vary that connote the spiritual act. On the contrary, on the surface it proves possible to locate the maximum of objective expansion (the maximum of the being against) in which consists the pure object or simple screen without consciousness—the only thing that remains of the object after having eliminated the spiritual movement. Both of these metaphysical levels of our concrete experience, conceivable only in the abstract, for Bergson, are not in themselves nor are they found in the guise of independent entities. Objectivity (objective screen) and intentionality (movement towards the object) combine, coming to constitute, only in virtue of this unity, the concrete experience in which in the most superficial layers there will prevail the character of the object, while in the deepest levels of reality, there will prevail that of the subject. Thus, following the lines of research laid down by Bergson, after the duration has been identified in the *Essai* with real time, starting from *Matière et mémoire* there comes to be outlined a new type of dynamic based on the various rhythms or levels of duration, as if they were various times, assimilating them in different forms of horizontal movements—the vertical movements consisting in a passage from one rhythm to another, that is, from one ontological level to another (doctrine of the multiplicity of the levels of being). In reality, from what has been shown thus far it appears clear how for Bergson time cannot be identified with any purely horizontal movement nor with a pure vertical activity, but must result from both.

3 The Conception of Unique Time and the Comparison with Einstein

Though the new developments of Bergsonian thought highlighted up until now certainly require a revision of the concept of time with regard to that delineated in the *Essai*, such a revision in an explicit form cannot be found in the French

---

[4]In the 1908 essay Bergson shows how the passage from perception to memory does not happen in the dimension of before and after, that is, in the proper dimension of duration: the memory, in fact, begins to exist contemporaneously to the present. In this guise, by saying that memory is formed contemporaneously to perception, one does not move in a horizontal direction (temporal succession), but in a perpendicular dimension (vertical movement): the past, then, is not what remains behind, but what comes out of time in a different dimension.

philosopher: the monistic concept of duration, in fact, remains alongside new developments without being merged with them. There happens here, in our judgment, something similar to what Jacob will define as *evolutionary tinkering*, that is, the theory according to which biological evolution occurs by reusing in the best way the material at its disposition present in the genome and not according to an activity that starts from zero to create new functions as in the case of engineering activities (Jacob 1977). Therefore, just as in the context of biology a new function is carried out by pre-existing molecules, slightly modified, or by new combinations of pre-existing proteins that according to the "environmental meaning" can develop teleonomic performances of a renewed character that will then gradually be established while evolution continues in other directions, in the Bergsonian philosophy the monistic concept of duration, if one will allow us the comparison, continues to reproduce itself alongside the development of new concepts like the bacteria *Escherichia coli* continue to reproduce inside and alongside *Homo sapiens*. It involves, in other words, conserving the profound and original things the *Essai* affirmed, avoiding making of time a second homogeneous and measurable means alongside space without, however, eliminating from time every relation with space (Szendrei 1989). Now, precisely to the relation between time and space Bergson dedicates the volume of 1922 entitled: *Durée et simultanéité*, in which the problem of time is taken up again explicitly through the close comparison with Einstein's theory of relativity. The time of the universe presented in this work is neither the pure and simple time of consciousness nor the diluted level of duration of extension that, in *Matière et mémoire* and similar essays, occurs parallel to the most tense duration. It is instead a single time that puts conscious interiority in contact with an "external field of experience" (Bergson 1922: 58–59).[5] The relation between external and internal is here mediated by the key concept of simultaneity that allows to make a determinate psychological moment correspond to a determinate situation of bodies in space, thus establishing a real link between internal and external without which it would be impossible to be able to speak of simultaneity since, as the French philosopher notes, also the simultaneity of the position of two or more exterior bodies cannot, ultimately, but be observed psychologically. Conscious of the incongruence of the notion of "simultaneity" with his old concept of time, Bergson introduces simultaneity by starting from the "simultaneity of flows" that consciousness can perceive together like single flowing (single act of intention) or separately by distinguishing them for their entire length (an action that is distributed

[5]Bergson, with a reasoning by analogy that allows one to see how in this work time places the internal in contact with the external, starts from the non-demonstrated presupposition that all consciousnesses perceive and live the same duration and imagines an infinity of such consciousnesses disseminated throughout the universe at such a distance that two consecutive consciousnesses, whichever they may be, share the extreme portion of their external field of experience. Because each of the two external experiences participates in the duration of each of the consciousnesses that in turn have the same rhythm of duration, the two fields of experience will have the same duration. Therefore, a single duration gathers up along the way the events of the totality of the material world (Bergson 1922).

but not divided in two) (Bergson 1922: 66). Though a rigorous immanentistic notion of duration excludes a simultaneity of flows no less than of instants which could at most be tracked down by an external observation, the French scholar admits, to avoid coming out of duration as flow, that the attention can allow the subject to observe the simultaneity of two or more different flows in virtue of the singular privilege of being "one and multiple together": the formation of the idea of instant or of that of simultaneous instants occurs naturally at the same pace as the acquisition of the habit of "converting time into space" (Bergson 1922: 68). As in *Matière et mémoire* it was considered illusory to attribute instants to time, so also in *Durée et simultanéité* it is declared that this would not be possible for without the tendency to apply the movement against the space travelled, causing the trajectory to coincide with the travelled path.

Given this, then, the introduction of the concept of simultaneity, sacrificing the characteristics of pure duration of the *Essai*, allows Bergson to grip Einstein's time[6] since this time can be counted only in virtue of the simultaneity of the instant: behold, therefore, the appearance of the uniqueness of a time that in the previous works would have been considered illusory. Of a real, psychological time and yet one that is measurable since it is "ascribed with a length" (Bergson 1922: 193).

On April 6, 1922 at the Sorbonne of Paris there took place the famous debate regarding the nature of time between Bergson, Einstein, and other philosophers and scientists that represents still now a milestone in this field (Bergson et al. 1922). Generally, scholars agree on the fact that, despite Bergson having certainly misunderstood many important points of Einstein's theory, his ideas deserve to be considered much more attentively than physicists usually do.[7] In agreement with this position we will continue the present examination by briefly fixing the terms of the comparison between Einstein (1905, 1916) and Bergson (1924a, b, 1934), a comparison that, though indirectly, can in any case contribute to introducing the new conception of time of the French philosopher. At the basis of Bergson's entire argument is the conviction that lived time, consistent in the effective duration of the

[6]The theory that Bergson takes into examination is restricted or special Relativity, that is, the hypothesis that Einstein (1905) elaborated to reconcile the relativity of straight-line uniform movement of classical mechanics with the invariance of the speed of light from the experiment of Michelson and Morley(Fano and Tassani 2002). This hypothesis implies that the measurement of the length of a segment referred to a system in movement is less than if it is referred to a system at rest and that, moreover, the measurements of time of any phenomenon are greater if referred to a system in movement than if referred to a system at rest (Pais 1982; Dorato 2013).

[7]Bergson received numerous critiques from other physicists supporting relativity, in particular from J. Becquerel and A. Metz, to whom he replied with three brief writings that were added to the second addition of *Dureé et simultanéité* under the form of *Appendixes* and later, in 1924, with an article and a letter in the *Revue de Philosophie*. In the end, however, he was convinced that it was not possible to come to an understanding, to the point that in 1926, to avoid further polemics and mistakes, he decided not to authorize new translations and later not even the simple reprinting of the book, even if this does not mean, as individuals sometimes read it, that he had given up his ideas, which on the contrary he reaffirmed again in 1934, although only in passing, but with unchanged conviction, in *La pensée et le mouvant*.

consciousness, should be accurately distinguished from spatialized time, that is, the time marked by the clock, because the latter is the fruit of an intellectual operation that reduces to measurable relations that which in reality is incommensurable. In consequence, simply measured time is for the French scholar only fictitious, while real time always requires the presence of an observer in flesh and blood to experience it (Bergson 1924a). For this reason in his estimation relativity demonstrates not the existence of different times according to the state of motion of each observer, but that of a single time, equal for all the observers, since the effects of the slowing-down of time are only fictitious since they are simply calculated and referred to but not experienced by anyone; on the other hand, when one assumes the point of view of the observer, supposed before to be in movement, this causes him now to have to be considered immobile, so that the slowing down disappears and for him time flows exactly in the same way with regard to the observer who had been supposed to be immobile at first (Musso and Paolo 2011; Cavilini and Musso 2012). Examining the theory of relativity (in particular, the restricted or special one), the French philosopher thus takes into consideration the possibility of a measurable time that allows a link between internal and external, committing the error, however, of distinguishing between real measurements and fictitious measurements: he sees only the possibility that two objects (AB, A′B′) in reciprocal movement are referred each to its own system (S, S′), not that only one of them (AB) is referred now to one system (S) and now to the other (S′). In other words, Bergson confuses the measured objects (AB, A′B′) with the systems of reference (S, S′) and does not take into account that, since there are two systems and objects, there are three possibilities: a) refer both the objects to S; b) refer both objects to S′ and c) refer one object to S and the other to S′. Bergson's error, thus, is above all a logical oversight, that is to say, the application of a reasoning that in itself is correct to a different object than that for which it is valid. In other words, the French scholar, while reasoning correctly, does not notice the fact that his hypothesis of the equality of lengths (according to which AB referred to S is equal to A′B′ referred to S′) is not in contradiction with Einstein's hypothesis of the equality of lengths with respect to S, but that it is simply different. The Bergsonian hypothesis, in fact, does not imply the equality between AB and A′B′ even if both are referred to S, or both to S′, nor that, if they are equal with respect to S, they must be equal with respect to S′. The French philosopher, therefore, does not detect the difference between saying AB really equals A′B′ (attributing each length to its own different system of reference) and saying AB equals A′B′ referring both to a single system of reference.[8] If he had noticed this, he probably would have realized that from the hypothesis of the

[8]The same error is repeated in the discussion of Einstein's example of the train aimed at showing that the simultaneity of two distant events is relative and in that of the "trip in a cannonball". In the latter case Bergson would have reason to deny that the trip causes one to become younger, because, in Einstein's paradox of the differently-aged twins, the difference is not due to the trip, but is introduced surreptitiously with the inversion of direction of which it is said that it should not be taken into consideration. The French philosopher, however, again uses an argument vitiated by the preceding logical error which for that matter is not even noticed by many of his later critics.

identity of content or of the interchangeability of the systems one deduces nothing that contrasts with the considerations of Einstein, whatever value those considerations may have (Mathieu 1954).

However much Bergson with regard to the German scientist may have committed other significant errors, some of which uninfluential (as for example the hypothesis of the microbe scientists or having overlooked the fact that psychological simultaneity is made possible by physical simultaneity and therefore the former adds nothing to the latter except the becoming conscious of it), while others were decisive, such as the pretention of discussing the problem of time solely in terms of strict relativity,[9] he is surprisingly right, even if for reasons different from those he offers, when he sustains that the belief in the reality of multiple times implies the belief in an absolute system of reference. In effect, although it is possible to establish the reality of two different times without for this reason having to admit that one of them belongs to a single system of reference, it is equally true that, to not fall into an infinite regress, sooner or later one will have to reach a system whose time has not slowed down with respect to that of any other system, a system, that is to say, that should be authentically inertial (a system in absolute rest or in absolute uniform rectilinear movement). Well, today we know that on the basis of the Big Bang theory there exist two systems of reference that are coherent with space itself (and in this sense absolute): one is the system of the galaxy on a vast scale, because in this case their intrinsic motion becomes negligible with respect to the movement of recession; the other is that constituted by the fossil radiation of the Big Bang, which fills in an almost perfectly homogeneous manner the entirety of space and which allowed Smoot (1993) in 1977 to measure the absolute velocity of Earth with an experiment he called "the new experiment of the shift of ether" with implicit reference to that preceedingly carried out by Michelson and Morley.[10] The point about which Bergson is correct is in sustaining that the subjective experience of time reflects a constitutive and ineliminable characteristic of temporality that impedes one from reducing it to a dimension entirely indistinguishable from spatial dimensions. Although Bergson erred in presupposing that time spatialized by physics is only calculated and cannot in any way be the object of experience, since duration is in effect a succession of states of consciousness, these states, however, are always based, in one way or another (at least in the case of human beings), on determined successions of events, that is, on something that happens inside space, but which at the same time proves to be able to be

[9]It is not possible here to analyze in detail the whole Bergsonian argument. For a more in-depth treatment of this question we refer the reader to Civilini and Musso (2012: 119–124).

[10]This moreover does not contradict relativity because the fact of placing oneself from the point of view of such an absolute reference does not give any practical advantage: the transformations of Lorentz, in fact, apply to it exactly as with any other system; in other terms, the cosmological system of reference is *absolute* but not *privileged*. Moreover, the space of relativity can be considered in a certain sense the "true" ether, since it is certainly "something" and not pure nothingness. What *truly* died definitively with the theory of relativity is only the *mechanical* ether (Civilini and Musso 2012: 123–124).

experienced only through consciousness understood as "living memory of change" or "form of succession" (Bergson 1907, 1922). So much so that in situations of extreme sensorial deprivation, the sense of time is greatly altered, while on the other hand that which is left of it continues in any case to be based on events, even if in this case they will be those of the internal vital rhythms of our body, that no matter how much they slow down and are reduced to minimal terms can never be completely suppressed without suppressing with it also the deepest root of our being: the pure subjective act of taking together different moments (transforming them from external into successive). It is precisely by starting from here that it is possible, in our consideration, to attempt to reconstruct Bergsonian time understood as "being endowed with a length" without being reduced to it (Bergson 1922: 272). Time and space thus once more reunite not because time becomes one of the dimensions of space, but, on the contrary, because space becomes one of the dimensions of time in the sense that it is the material itself of which time is formed. If, then, when interpreting the theory of Einstein the distinction between real and fictitious times invalidates on the logical level a large part of the reasonings of *Durée et simultanéité* (think, for example, of the paradox of the twins and of general relativity), when interpreting Bergsonian thought it instead proves to be a clarifying element. We cannot but notice, for example, how, unlike the *Essai* where Einstein's times would be considered all fictitious, in *Durée et simultanéité* it is possible to speak, in agreement with Bergson, of a time that is at once "real" (psychological) and "measurable" (mathematical) precisely in the sense of lived reality, unlike the "elongated time" that is only mathematical (Bergson 1922: 173). Time will thus have two components (as long as one understands composition in an abstract sense and not as mixture or juxtaposition): one formal, which is the "taking together", and the other material, which is space itself, which is taken together and forms the material of time. In other words, it involves two immanent dimensions in the single temporal dimension which therefore comes to be "curved" and dynamic.[11] Here then we see the delineation of a time that is configured like a directly perceivable concrete flow (not relative but absolute) through the presence in it of two moments that, immanent in the single dimension of time, give rise to the curved and absolute dynamicity of the temporal succession: space (represented by the tangential movement), on the one hand, and the subjective act of taking together or intentionality (represented by centripetal force), on the other. Time, therefore, in agreement with the second Bergson, is made of space as of its material in the same sense in which the curved movement is made point by point by a movement in the direction of the tangent and is at the same time in each of its points referred to the subject in the same sense in which the curved movement is derived continuously towards a center. In this guise, it becomes possible to infer, as Mathieu (1954) rightly notes, that time contains in each instant the possibility of space since space

[11]This single time that lasts, according to Bergson, is that which the relavitistic physicists call "proper duration of a phenomenon" and which, unlike the others that are reduced to pure and simple lengths, is a time without doubt endowed with the length that measures it.

consists in that abstraction that time becomes whenever, in any instant, one does without the reference to a subjective center, that is, from that intentional act able to make successive the moments that in themselves are only exterior. Thus, time and space come to coincide in the instant (instant by instant time is space) while space is new at each instant just as the curved movement has, point by point, the direction of the tangent and the tangent is new at each point (time is curved and concrete, while space is flat and abstract).

In the light of all this, therefore, we can infer, with Bergson, that space is "the abstract tangent of time" which it is not necessary to think of in any case as "psychological" (Bergson 1922: 127–146). The reference to the center, in fact, is not added externally to the objectivity of time, but remaining immanent to it confers upon it a different form: if from the material point of view time continues to be exteriority of instant by instant (from spatial point to point), from the formal point of view it becomes, instead, a concrete and absolute flowing, a form, that is to say, which while being immanent to the objective dimension of time is not reduced in any case to it (Bergson 1922, 1934). All this allows us to offer two considerations. The first is that while one can always pass from time to space, it is not instead possible to recompose time by summing up spaces or instants; on this point one cannot agree with Bergson who denies the fact of being able to construct the curvature of time with infinitesimal tangential elements. Although this reconstruction of time by integration is impossible as a real construction, it instead subsists as an opportunity of intellectual reconstruction especially at the level of the time of science, that is, of the time made of instants that are taken from concrete time. The second consideration is directly linked to the first and refers to the fact that Bergsonian duration, introducing itself in the real world, impedes it from reducing itself without deformation in the deterministic and spatializing schemas of science. The time of the procedures of science, in fact, as we have just now hinted, in reality is identifiable with space since in order to establish pure objectivity, eliminating all that is subjective, such procedures overlook the continuous reference to the intentional center, causing the curve of time to coincide with the tangent (Mathieu 1954).

In addition, such considerations allow us also to recognize in Bergson the merit of having anticipated with foresight some methodological shifts that in the scientific realm would be verified only many years after his intuitions, like for example, the renunciation by the natural sciences to place reality inside a single deterministic schema (Prigogine and Stengers 1988). Not unlike Augustine, still today we do not know what time is but it is to the complete definition of Aristotle that, after the discovery in physics of complex systems and of the laws of chaos, we can lead back the laws of motion (Prigogine and Stengers 1979). The intrinsic measurement of motion imposes the perspective of a before and after. Motion conceived by Galileo and his successors articulated the instant and eternity. In every instant, the system was defined by a state that contained the truth of its past and of its future. Motion as we conceive of it today gives a width to the instant and links it to becoming. Every instantaneous state is a memory of a past that permits one to define only a limited future, circumscribed by an intrinsic temporal horizon. The definition of the instantaneous state thus breaks the symmetry between past and future and the laws of its evolution multiply this breakage of symmetry (Szendrei 1989). Bergson, to express the solidarity that unites us to the time of things had written that "we must wait for the sugar to dissolve" (1907). It is this solidarity between our time and that of phenomena that is implicity translated by the probabilistic laws that allow us to foresee but not to reconstruct the past. It is this solidarity that the dynamics of chaotic systems affirm in an explicit manner and that in recent years in the context of complexity theory, has been developing in an ever more articulate manner, confirming albeit indirectly the brilliant philosophical intuitions of Bergson. At the level of processes of semantic categorization, for example, today the procedures of unification, in agreement with conceptual bonds, seem to nestle and find their foundation in modules and attractors that operate at the level of patterns of connection as well as (organic) instruments-systems of measurement that come to multiply by degrees. Behold an autonomous selective production of forms that, since it is modulated by concepts and linked and connected through the *telos*, comes, in the end, to make itself into vision through principles, production of forms (natural modules of connection animated by an internal code) able to articulate itself in agreement with a precise intelligence, one that unifies it in time (Carsetti 2004). Hence the very possibility of the "presentation" of an originary meaning that contemporaneously deploys itself and divides itself with the confines of a "work-form", where the procedures of reference come to be delineated in an entirely special way: it can be referred to the real, not only through categorial intuition, through filling by way of the intuitions and the construction of the Bergsonian "tangle", but also through intuitive categorization, through the emergence of concepts within the meanderings of intuition inasmuch as they are concepts filled with meaning.[12] Here is the first source of that continuous "addition" of new nuclei of creativity that characterizes the logic of the living being of which Bergson spoke (1903, 1907, 1934) and that in *Durée et simultanéité* allows to configure the intuition of duration with valences that were surely extraneous to the *Essai*, reinforcing moreover the intuition itself through the use of analogies and

[12]Confirming some Bergsonian intuitions, the contemporary theory of self-organization has shown how forms are articulated by concepts. They can do so on the basis of the inscription of thought in the determinations of time (the rhythm-scanning operated by the form production), i.e. through linkage by "ring-threading *via* schemata". The schema is the "reduction-medium" that allows the unification of the forms on generative bases, and therefore, by concepts. Hence the necessity of a continuous connection between processes of "rational perception" and "processes of intuitive categorization" through recourse to a process of self-organization that allows the inspection and overcoming of limits, as well as the possible invention of new procedures. It is that, therefore, which allows the birth of the cognitive activity and the generation of languages in continuous evolution (Carsetti 2004, 2013). Thus, the concepts, the attractors come to live and attune themselves in a dynamic and co-evolutionary context: that regarding choices and fusions that arise from the process of production of the forms on the basis of specific procedures of inscription and of a operative nestling that Bergson represented with the images of the overturned cone and of the curvature of time capable of holding together the formal element of time (the curve) inside a material element (the line).

inferences that were then excluded or avoided. The progressive concentration of duration little by little that descends the levels of being offers us thus that duration-principle concentrated in eternity that does not coincide with our experience but which our experience needs. The more we sink into real duration, Bergson sustains in *Introduction à la métaphysique* (1903), the more we place ourselves again in the direction of the principle, which moreover is transcendent, which we participate in, and whose eternity does not have to be an eternity of immutability, but an "eternity of life" (Bergson 1903: 176).

To find, therefore, the authentic Bergsonian time we must compose the doctrine of the various horizontal durations with different rhythms (Bergson 1889, 1903) and that of the vertical dynamism or subjective effort (Bergson 1902, 1908) that passes through the levels of consciousness. These two movements, in fact, in agreement with the works of 1922 and of 1934, allow themselves to be isolated by abstraction from the curved dynamicity of time. This explains how the French philosopher can speak at the same time of durations in the plural (1903) and defend in the volume of 1922 the unicity of universal time, of a single time that offers the possibility of being considered in infinite different ways because it contains in itself, in a single dimension, an objective material aspect and a subjective formal one or one of a living intentionality.

4 Living Time

Up to here, by facing the problem of time in the thought of Bergson, we have seen how the evolution by *do-it-yourself* of the concept of duration favors new developments of the philosophy of intuition. Duration, which in the *Essai* connotes the experience of a non-measurable lived experience, while in *Matière et mémoire* takes on rhythms of differing intensities to justify the relation between perception and memory, in *Durée et simultanéité* is posited not only as a criterion to discern what is real from what is artificial, but also to legitimize the measurement of reality, that is to say, to reestablish that point of contact between time as duration and space, that formerly seemed compromised by the loss of ontological consistency of extension as a dimension of the corporeal. At this level, therefore, in which space comes to be outlined as the abstract tangent of time, it becomes possible to examine the emergence of concepts within the labyrinth of intuition as concepts imbued with meaning: here is the first source of that continuous "addition" of new nuclei of creativity that characterizes logic itself of the Bergsonian living creature where that which, as an ideal relation, is time, as a real relation, becomes life, that is, continuous opening of a register in which time is inscribed, thus becoming flesh in the very nature of every organism (Bergson 1907). Organic life, in fact, cannot but develop temporally, since vital properties are never entirely realized, but are always in a process of realization: the becoming of every living being (its potential richness) is made possible by the fact of never being definitively complete. Therefore, finding ourselves before an analogous situation to that we experience in our duration, understood in Bergsonian terms as the authentic increase of being that is realized in the co-penetration of elements that are always qualitatively different (Bergson 1902, 1903), it is necessary to attribute a duration also to the living organism whose present is always more than his past. Thus, by analyzing life it proves possible to detect the same components identified in time: a multiplicity of external parts and the reference to a certain grade of recollecting into a unity not only the spatial exteriority of the parts of the body, but also that temporal exteriority of the successive moments in which there surfaces a unitary orientation that surpasses temporal distance.

In *Évolution créatrice*, where duration is judged as the very material of the real, Bergson shows how that which for time was the spatial reference to a center, for life is the immanence of a principle vital to the body, of a principle, that is to say, which, while eluding on a purely objective level the forceps of the anatomist, on the individual level constitutes the organism itself, distinguishing itself from the body with which it places itself in a relation similar to that which constituted time. The phenomenological analysis of time, therefore, can serve to illuminate the study of life. In the measure in which the French philosopher conserves of time the monistic conception of the *Essai*, his doctrine of life, too, is conditioned by it. However, as from the theory of the multiplicity of levels there is a way to develop a different conception of time, so the making extrinsic of the monistic thrust does not exhaust all the content of the *Évolution créatrice*, which, in agreement with the interpretation offered by Mathieu (1954), understands life as immediate presence of the principal that is vital to the organism and not as the identity of the organism with its principle: the metaphysical distinction of the profound and superficial levels or modes of being allows, thus, to grasp a new and more authentic immanence profoundly different from a monistic type identity.

To this end it is opportune to reflect on the hinge around which oscillates the entire Bergsonian opus, that is, the problem of individuation. The appearance of individuality, in fact, is sufficient to distance the interpretation that *Évolution créatrice* gives of life from the monistic conception of the pure duration of which evolution might seem nothing other than a cosmic extension. Life tends to reabsorb living beings into itself, while not going so far as to do so, while pure duration does not admit within itself individual nuclei. The individuality of the organism, in brief, reveals however an immanence different from that of monism which characterized pure duration as the emphatic negation of every spatiality. The thrust is immanent and immediately present in the living body but one should avoid transforming this immanence into an identity that reabsorbs everything into the *élan vital*. Given this, then, we can infer, with Bergson, that life is present only where there is a certain individuality that is always in some way the presence of a unitary principle to a plurality of elements made organic by it. The individuation of such an organicity (incarnate time) is an effect both of matter and of what life bears within it (Bergson 1907). Vital immanence is thus immanence in another, presence of something that transcends this other. It is precisely the manifestation of a profound unitary principle (with regard to the materiality of the body) on the level of objectivity that allows us to detect in vital reality the presence of a "form" that constantly transcends space. It involves, in other words, an attempt at synthesis and of an enlargement of the prospectives preceedingly elaborated, that goes far beyond Spencerian theory. The entire dimension of biological life, like that of consciousness, is liberated from any determinism to place itself on a level of unpredictability and the *élan vital*, perhaps the most famous expression of all of Bergson's thought, expresses an idea of life as continuous creation, to the point that the same inorganic world, matter, would represent nothing other than a momentary halt in that thrust, of that dynamic thrust that is fragmented in different individualities in contact with an originary matter that is not identical to the concrete matter that is posterior, and not anterior, to life (Bergson 1907).

References

Bergson, H. (1889). *Essai sur les données immédiates de la conscience*. Paris: Alcan.
Bergson, H. (1896). *Matière et mémoire*. Paris: Alcan.
Bergson, H. (1902). L'effort intellectuel. *Revue Philosophique de la France et de l'étranger, 53*(1), 1–27.
Bergson, H. (1903). Introduction à la Métaphysique. *Revue de Métaphysique et de Morale, 11*(1), 1–36.
Bergson, H. (1907). *L'évolution créatrice*. Paris: Alcan.
Bergson, H. (1908). Le souvenir du présent et la fausse reconnaissance. *Revue Philosophique de la France et de l'étranger, 66*, 561–593.
Bergson., H. (1922). *Durée et simultanéité. À propos de la théorie d'Einstein* (2nd edn.) [also contains the Appendices]. Paris: Alcan.
Bergson, H. (1924a). Les temps fictis et les temps reél. *Revue Philosophique de la France et de l'étranger, 24*, 241–260.
Bergson, H. (1924b). Réponse de Henri Bergson. *Revue Philosophique de la France et de l'étranger, 24*, 440.
Bergson, H. (1934). *La pensée et le mouvant*. Paris: Alcan.
Bergson, H., et al. (1922). Séance du 6 avril 1922. La théorie de la relativité. *Bullettin de la Société française de Philosophie, 22*, 102–113.
Carsetti., A. (2004). Seeing, thinking and knowing. Meaning and self-organisation. In A. Carsetti (ed.), *Visual cognition and thought* (pp. 307–330). Dordrecht: Kluwer.
Carsetti, A. (2013). *Epistemic complexity and knowledge construction*. Berlin: Springer.
Civilini, E., & Musso, P. (2012). Il concetto di tempo nel dibattito tra Einstein e Bergson. *Epistemologia, 35*, 112–129.
Dorato, M. (2013). *Che cos'è il tempo? Einstein, Gödel e l'esperienza comune*. Roma: Carocci.
Einstein, A. (1905). Über die Elektrodynamik bewegter Körper. *Annalen der Physik, 4*(17), 891–921.
Einstein, A. (1916). Die Grundlagen der allgemeinen Relativitätstheorie. *Annalen der Physik, 4* (49), 769–822.
Fano, V., & Tassani, I. (2002). *L'orologio di Einstein*. Clueb: La riflessione filosofica sul tempo della fisica. Bologna.
Jacob, F. (1977). Evolution and tinkering. Science 196 (4295): 1161-1166.
Mathieu, V. (1954). *Bergson. Il profondo e la sua espressione*. Torino: Edizioni di "Filosofia".
Musso., P. (2011). *La scienza e l'idea di ragione. Scienza, filosofia e religione da Galileo ai buchi neri e oltre*. Milano-Udine: Mimesis.
Prigogine, I., & Stengers, I. (1979). *La Nouvelle Alliance. Métamorphose de la science*. Paris: Gallimard.

Prigogine, I., & Stengers, I. (1988). *Entre le temps et l'éternité*. Paris: Librairie Arthème Fayard.
Pais, A. (1982). *'Subtle is the Lord...'—The science and the life of Albert Einstein*. Oxford: Oxford University Press.
Pessina., A. (1987). Le illusioni della coscienza e l'intuizione come metodo. L'impostazione dell'"Essai" di Henri Bergson. *Rivista di filosofia neo-scolastica, 79*(1), 96–117.
Pessina, A. (1994). *Introduzione a Bergson*. Roma-Bari: Laterza.
Smoot, G. (1993). *Wrinkles in time*. New York: William Morrow & Co.
Szendrei, E. V. (1989). Bergson, Prigogine and the rediscovery of time. *Process Studies, 18*(3), 181–193.
Taroni., P. (1999). *Tempo interiore, tempo oggettivo. Bergson e Piaget. Il concetto di tempo dalla filosofia della vita all'epistemologia genetica*. Urbino: Quattroventi.

Concepts of Time in Husserl

Felice Masi

Abstract Temporality represents the most important and difficult question of phenomenology: decisive for its idea of phenomenon and consciousness. What means that time is the appearing itself, so not a time of consciousness but the consciousness itself: this is the phenomenological question about the origin of time. Composed in three decades approximately—from 1904 to 1934—Husserlian contributions phenomenology of temporality constitutes the most extensive *corpus* about this matter in the canon of occidental philosophy. They lead in three main directions and correspond to the same number of periods of their development: (a) the *mathesis* of intentional manifolds (1904–1911), the metaphysics of individuality (1917–1918), the theory of temporal self-constitution (1929–1934). After the description of the phases, the sources and the internal articulations, the paper makes room for a brief and essential glossary of phenomenology of temporality, made up of some of the most considerable and aporetic notions: the retention, and its bond with protention, individuality and its elusive essence, the flow and the stream. Lastly, the paper inspects and examines some of the most remarkable critics to phenomenology of temporality, from Heidegger to Derrida, from Bergmann to Lévinas, in order to demonstrate how leading was its role in the whole philosophy of the twentieth century.

The most important question, and at the same time that which fate sometimes relieves us from having to answer (Husserl 1913: 194; Husserl 1928: 346): so time appears in the Husserlian phenomenology. And it is precisely the crossroads between the ambition to resolve its enigmas and the fear that any effort may prove to be vain that describes in the best of ways the motive of that uninterrupted rewriting to which Husserl subjected his reflections on time, until composing, in three decades, from 1904 to 1934, the largest *corpus* that had been dedicated to this topic in the entire history of Western philosophy.

F. Masi (✉)
Philosophy Section, Department of Humanities, University of Naples Federico II, via Porta di Massa 1, 80133 Naples, Italy
e-mail: masi.felice@libero.it

F. Santoianni (ed.), *The Concept of Time in Early Twentieth-Century Philosophy*, Studies in Applied Philosophy, Epistemology and Rational Ethics 24, DOI 10.1007/978-3-319-24895-0_7

1 History and Chronology

Conventionally three phases are proposed, corresponding to as many collections, not always organic, of manuscripts, notes, lectures:

(a) the first, having as a nucleus the fourth part of the course of lectures of 1904–05—dedicated to the *Fundamental features of phenomenology and theory of knowledge*—subjected in the following years to ample revisions, handed over in 1917 to Edith Stein so that she would prepare a version for the book edition published later in 1928, edited by Martin Heidegger in the *Jahrbuch*, with the title *Lessons on the Phenomenology of the Internal Consciousness of Time* (hereafter PhZb), and finally edited in 1966 by Rudolf Boehm as volume X of the Husserliana (hereafter Hua)[1]
(b) the second redacted between 1917 and 1918, during two periods of vacation in Bernau, in correspondence with the beginning of Stein's redactional work and in the attempt to integrate the preceding analyses, and then taken up in 1928 at the care of his young Freiburg assistant Eugen Fink, in sight of the publication of what was supposed to be his *Hauptwerk*: this was conserved for a long time in the Finkian personal library and only in 1969 transferred to the Husserl-Archiv of Louvain, to then be published in 2001 with the title *Die Bernauer Manuskripte über Zeitbewusstsein* (henceforth BM), as volume XXXIII of the Hua;
(c) the third, composed between 1929 and 1934, as the second final part of the volume that in that years was supposed to have been prepared by Fink—aimed at a *renewal of the principles of rational metaphysics*[2]—was interrupted with the beginning of the drafting of *Krisis* and with the proposal of the project of an enlarged German edition of the *Cartesian Meditations*: his manuscripts were published in 2006 as *Späte Texte über Zeitkonstitution (1929–1934). Die C-Manuskripte* (henceforth ST), as volume VIII of the *Materialen* of the Hua.

---

[1]Showing the stylistic and conceptual deformity of the original text prepared by Edith Stein, drawing from materials of very different times and sometimes forcedly homogenizing them, the originals of which, moreover, were lost, Boehm's edition was further emended by Bernet (1985), regarding the dating of some texts included in part B of Hua X, and recomposed in four groups: the first (Tx. n. 1-17) dating back to 1893–1901, the second (Tx. n. 18-35) to 1904–1905, the third (Tx. n. 39-47 and 51-52) to the period between the winter semester 1906/07 and the end of August 1909, and the fourth (Tx. n. 48-50 and 53-54) to 1909–1911. Therefore the re-dating regards the texts n. 18 (from 1901 to 1904), 48-50 (from 1908 to 1909) an 36-38 (from 1909 to 1917). Particularly relevant is the postdating of these last three texts that constitute the second part of the *Seefeld Manuscripts* (Bernet 1985: XXXV; Husserl 1985: 283 ff.)– following the Tx. n. 35, dated already by Boehm to 1905, according to a title given by Husserl himself "Seefeld manuscripts and older manuscripts on individuation. Seefeld 1905. Individuation" (Husserl 1928: 244)—since they would prove contemporary to BM, with regard to which they exhibit also a clear theoretical coherency, represented by the topic of temporal individuality.

[2]Letter of Husserl to Heidegger of 03/28/1918, in Hua, Dokumenten, III, IV, p. 130; see Bernet, R.-Lohmar, D., *Einleitung* in (Husserl 2001), p. XXII.

Already from this first historiographical synthesis, it is clear what should have constituted the interminable effort that seemed to begin anew each time from the start with the aporias and the semblance of their coming resolution: each time that the reflections on time seemed to merit a completed exposition, giving way thus to the editorial work of others, both in 1917 and in 1928, their rethinking opened another road, from which point of view that precedingly taken proved imprecise, incomplete or simply too brief.

2 Sources

The premise of the phenomenology of time—or of *the question about the "origin" of time* (Husserl 1928: 9)—can be expressed as a double lack that is outlined in the end notes of the *Logical Investigations* and to which Husserl himself gives the common title of *Zeit und Erscheinung*: the impossibility to describe the differences between perceived, remembered and imagined with just the instruments of the variations of quality of act, too coherent with the Brentanian theory of time. However, the task thus outlined can only come from an analysis of intentional objectuality (of time) and its inclusion in a wider transcendental reflection that traces back to the definitions of 'I' and of subjectivity.

The roots of the stalemate of the first years in Gottinga and the sources to which its overcoming appeals can be schematically summarized as follows:

(1) the Brentanian doctrine of the originary association between perception and memorative modification and more generally the consequent Aristotelian-empiricist recourse to the imagination to make up for the defects of representative adequacy both in the spatial and temporal givennness—both still widely present in the *Logical Investigations*;
(2) the limits of the *Fifth Investigation* to account for the unity of the manifestative discourse and of the specificity of the intentional object with regard to the distinction between *real* (object or physical or physiological process) and *reel* (psychic, immanent object, intentionally inexistent)—still in full accord with the magisterium of Brentano;
(3) the reading—attested already in the *Second Logical Investigation* (Husserl 1984: 200)—of the *Principles* of James (1890) and the isolation of the concept of *fringes*, that is, of the perceptual path traced by present perception, a pathway "weakly perceived" (James 1890: 275) or again not properly conscious, that is, of what PhZb would address in terms of phenomena of consciousness, which are not the object of primary observation (Husserl 1928: 155), fleeting sensations that pass away (*Übergangsempfindungen*), phenomena that change content;
(4) the news of the polemic that arose between Stern (1897) and Meinong (1899) and taken up by Husserl in 1904 regarding the attributes of punctuality and simultaneity of the apprehensional act; from this exchange there are taken

decisive terms such as *primäre oder frische Erinnerung*, or again *Präsenzzeit* and *Zeithof* (in which one still notes the Jamesian reading), as well as the sharing of the battle against "the dogma of the momentariness of the whole of consciousness" (Husserl 1928: 22), although the insoluble problem still remains of the alternative between a psychological study that does not make a problem of the intentionality of temporal objects and a theory of objects that remains mired in a murky collective doctrine of multiplicity and that prejudicially takes on temporal notions, such as that of precedence or the past (Husserl 1928: 23–25, 223–235);

(5) the recourse to the critical revision—initiated with a seminar in 1904—of the notion of *Retention* in the dual version presented by the *Essays* of Locke and by the *Nouveaux Essays* of Leibniz, from whom derives the distinction between the psychic performance (or faculty) of retention and the condition of continuity which each of the perceptive elements and their transpiring underlie, in which the not-yet-extinguished state of the preceding one, that is, its being *almost* extinguished, occurs together with the imposition of the successive one;

(6) the going-back to the Augustinian tradition according to a triple movement:

   (a) the reference to the Neoplatonic-Cartesian motive of the *epistrophé*, which serves as a tacit methodological premise of the analysis;
   (b) the rhetorical-apologetical recourse to the characteristics of discursive unspeakability and of the exceptional nature of temporal phenomena;
   (c) the centrality of the measuring function in the determining of the temporal dimensions.[3]

3 Meanings of Time

The chronological analysis does not, however, fully account for the overlapping between the different drafts nor for the theoretical articulation that can be deduced from the different phases. The internal story of the phenomenology of time does not coincide with its ideal profile (Brough 2010: 22–27; Zahavi 2011; Sokolowski 1974). Such a repeated treatment is already a characteristic proper to phenomenological temporality, its *periodicity*. It does not consist only in the sequence of different moments, but each of these moments, or rather, *periods*, is presented in itself as composed of different levels of inquiry. It is possible to see, indeed, how

[3]This last topic is that which most clearly distinguishes the two contemporary versions of that which prima facie could be understood as the common Augustinian tradition: the phenomenological version and the Bergsonian version (Ingarden 1922; McLure 2005; Schnell 2004).

the succession of the versions of a phenomenology of time corresponds to the privileged thematization of one of the levels of which it is composed.

Already in the *Fifth Logical Inquiry*—surveying the meanings of consciousness —Husserl identifies, albeit implicitly, some of those which will be the levels of his analysis of time:

(1) consciousness as succession and multiplicity of immanent lived experiences;
(2) consciousness as the unity of these lived experiences and specifically as an "internal" unity;
(3) consciousness as intentional lived experience (Husserl 1984: 353–363).

In the course of the investigation, Husserl, to be able to dedicate himself to the last meaning, as properly phenomenological, and analyze thus the components of the intentional content, that is, the material and quality, and their re-composition in the intentional essence, must leave aside not only the first two meanings, but also the determination of the sequence of sensible contents—excluded by principle from that level of phenomenological study since it corresponds to a genetic analysis, too close to, if not entirely coinciding with, the psychological study. Such a decision, inextricably linked to that of denying oneself the possibility of having recourse to the integration of any kind of notion of a pure I—as the internal unity of consciousness—other than the generic name, the nominal vicarious form, of each trait of the perceptions that are adequated, that is, of one's own perceptions and one's own inasmuch as they are present, will be made the object of a profound revision in the second edition, contemporaneous with the publication of the first volume of the *Idee*.

Still between 1909 and 1911—taking up the distinction made already in 1907 between two concepts of transcendence, a naive one that can be expressed as extraneousness or independence, and a phenomenological one, that corresponds to inadequacy, impropriety or incompleteness—Husserl will ask

> Now does not our thematising regard, by posing the enduring perception as enduring, transcend what alone is really given, the now, and along with it posit something that is no longer really given: the continuity of past perceptual nows? (Husserl 1928: 353).

exposing himself thus to a dual result: either such an intuition implies a transcendence, or a grade of non-saturation, a lacuna, a margin of emptiness; or instead it denies it, thus denying not only the ability to intuit the duration of that which is perceived, but also that of the perception itself, falling into an *absolute skepticism* (Husserl 1928: 354).

In this way it is possible to comprehend why retention and I appear together in the text of 1913 (Husserl 1984: 357–358), both invoked to make up for that which is lacking in an unsustainable notion of adequacy and perceptive propriety. None of the first four sources listed above—which integrally include also the primitive formulation of the apprehensional content-form schema (obtained with particular reference to the consideration of the Stern-Meinong polemic)—would have proved sufficient for the inauguration of a phenomenology of time which implies rather the thematisation of the nexus between the last two meanings of the *Fifth Investigation*, between unity and intentionality, and the recovery of what in it remain entirely excluded, the leading, sensible, pre-intentional content. And the first result of such a rethinking was precisely the coining—datable to 1907 (Bernet 1985: xxii, xxxvi–xlviii)—of the term retention, that on account of which "consciousness can be made into an object" (Husserl 1928: 123).

In PhZb that first rudimentary subdivision takes on the more comprehensive aspect of

(a) the absolute time-constituting flow of consciousness;
(b) the constituting multiplicities (*Mannigfaltigkeiten*) of appearance belonging to different levels, the immanent unities in pre-empirical time;
(c) the things of empirical experience in objective time (and here it would be necessary once more to distinguish different grades of empirical being, from the thing of the experience of the single subject to the thing of physics) (Husserl 1928: 77).

To this corresponds, while not coinciding, the articulation into

(a) "the absolute timeless consciousness", in which subjective time is constituted:
(b) the taking up of the perception as phenomenon: "I am directed towards the perception, the appearance and what appears in their correlation": the subjective time;
(c) "the perception of empirical objects in the ordinary sense: there they stand" (Husserl 1928: 116)

The last point could be said to be that in which "I have perception of a steam whistle or, rather, of whistling of the whistle" (Husserl 1928: 117). Afterwards we should mention the perception "of the tone-content itself that endures and of the tone-process in its duration, without regard to its insertion into nature" (Husserl 1928: 117), in the objective-real time (or also scientific-natural). This level is that of the varieties of apparition; while the taking up of the perception as phenomenon renders "the perception of the tone-now and at the same times heedfulness of the tone-just-having-been combined with it" (Husserl 1928: 117). From this derives, moreover, the possibility of perceiving the consciousness of time in the "now", that is, the "now" of the apparition and of its passing, but not the "now" of consciousness, so to speak, and of its passing.

Each "now" is in fact a reciprocal demarcation of change and non-change: is it however possible to thus label consciousness inasmuch as it is the *continuous flow of appearance-phases* (Husserl 1928: 118)? This exposes a *höchst Merkwürdiges*, as if to say the greatest of oddities, but also that which most deserves to be pointed out.

Flow, however, is only a *Bild*, a *Gleichnis* (Husserl 1928: 79), an image, an analogy to express that in which nothing lasts (Husserl 1928: 118), a metaphorical remedy to express that for which "we lack names" (Husserl 1928: 79), the *absolute subjectivity*, the constant alteration that *absurdly* "flows precisely as it flows and can flow neither 'faster' nor 'slower'" (Husserl 1928: 78). An alteration that appears as time and that self-appears reflexively as constituting time, but does not have its own time: it does not have nor is it time.

An ulterior and clearer subdivision, or at least more precise, is that between

(1) «the *appearing (flow)*;
(2) the *appearances*;
(3) the *appearing object*» (Husserl 1928: 368).

Or—inverting the order of the list drawn up by Husserl:

(1) *the one and only* (einzige) *flow of the running-off of time*;
(2) *the one immanent time*;
(3) *the unity of time-consituting consciousness* (Husserl 1928: 358).

Again in BM there are distinguished:

(1) "the originary essential peculiarities of originary life (*Urwesenegentümlichkeiten des Urlebens*)", "the originary continuum of impressionality or of each present time after time of life (*Urkontinuum der Impressionalität oder der jeweiligen Lebengegenwart*)" (Husserl 2001: 268);
(2) "the variety [that is constituted in 1)] of the lived experience of phenomenological time, the constant plurality of the temporal-phenomenological events that follow one another and coexist," (Husserl 2001: 268);
(3) in such "varieties there can in their turn be constituted the unities of an objectivity 'transcending consciousness', like the unity of spatial things, of animal beings, etc." (Husserl 2001: 268–269).

And even in ST there are listed:

(1) the flow of the originary-primordial temporalization, of self-temporalization;
(2) hyletic time, of the hyletic unities, of the passive-intentional constitution, of the things presented;
(3) worldly time, objective and shared, the historical time of experience and of its natural history (Husserl 2006: 80–85).

Finally such a composition can be appreciated in paragraph 22 of *Experience and Judgment* (Husserl 1948: 104), where a distinction is made between:

(1) the contemplative intuition, *before* exposition/explication/explicit making—or in the terms of PhZb: *die Erscheinung ohne Auffassung*, the appearance without apprehension (Husserl 1928: 111)—: the unobstructed exercise (*ungehemmte Auswirkung*) of perceptual interest;
(2) explicative exercise, intentioning individuation: internal horizon;
(3) explicative operation (*Leistung*) of the coprensence of the individuated (object) into the multiplicity of the individuals: external horizon.

It is evident how the phenomenology of time in its entire development preserves the articulations we have seen, although each of its *periods* centers on one of the levels of which it is composed, modifying the equilibrium to the point of inverting it, so as to conclude that:

(a) PhZb, addressing what we have always found in second place—that is immanent or pre-empirical or phansiological time or the variety of the *Erlebnisse* or hyletic time—defines itself as an analysis of the temporal multiplicity-modification, that is as a *Mathesis of variety*;
(b) BM, instead, which Husserl consecrates to the individualizing unifications of these varieties, to these things—here that flow, to the noematic nature of time and to its intentional character—thus founding what we have found in the first two positions—can be entitled (as for that matter the Author himself solicited to be done) a *Metaphysics of individuality*;
(c) finally, ST, dedicated to the analysis of self-temporalization (*Selbst-Zeitigung*) starting from a notion that had already appeared in 1917, that of "concrete living present", *strömend-verharrend*, flowing-permanent, and therefore to comprehend in the self-temporalization the proto-figuration of the being of transcendental subjectivity (Husserl 2006: 6; Held 1966)—in which there are founded the characteristics of the now, of the *nunc stans* and of the flow—could thus be defined as a *Preontics of time*, in which there finds space a real and proper phenomenological archeo-mythology[4]

4 Phenomenological-Temporal Vocabulary

If it is true that phenomenology is above all a new language of thought, it could be useful to redact a brief essential vocabulary of its treatment of time, formed by a few of the most decisive and complex terms.

(1) *Retention/Protension.* Expressions from two different epochs and demands, having arisen in Husserlian usage ten years apart from each other, retention (1907) and protension (1917) exhaust the possible varitions of the originary presentation, that is the noetic aspect of the phenomenological-temporal analysis. However, even when they are used in combination they do not lose the functional difference that separates them: constitution of the pre-whole *Jetzt-Gewesen* the first, temporal expression of the same intentionality—and thus of the expression-intuition, look-filling coupling—the second. If it is true that the main distinction is that regarding their intentional status, yet this characteristic—above all for retention—has not received a coherent definition in the development of Husserlian reflection, nor in the analyses of critical literature (Duval 1981; Boehm 1981; Bernet 1983; Brough 1989; Kortooms 2002; Mensch 2010). Assuming that retention is a phase of the consciousness-apprehension of "now", without its own extension—where the same "now" is a limit always linked to a retention, that is in turn the limit of a continuity of retentions—and that therefore it is a non-independent part, which follows—by

[4]Derrida speaks of phenomenological archeo-teleology (1972: 60).

> an a priori law—the "now"—any "now" that it may be: perceptive, memorative, imaginitive—it remains to be understood what kind of modification it represents. The retention of A is its "having been perceived": it is not now, but in the now is given its not-now. Therefore, retention cannot be a representative or symbolic or apprehensional modification.

> Then 1) T1 appears as past, as "lying back in time" in relation to T2; 2) and together with this the following is given with evidence, even if, perhaps, it can be grasped <only> by means of reflection: the perception of T1 is a perception that has been; it lies back in time in relation to the perception of T2 (Husserl 1928: 199 [Tx n. 21, 1904]).

In PhZb retention is thus defined as a peculiar intentionality, an intentionality with a specific character of its own (Husserl 1928: 33, 122), which should be described as an improper intentionality. Well, while in some texts, such as *Experience and Judgment* (Lohmar 1996) this attribution of intentionality is preserved—albeit in the terms of an *intentional modification in the realm of the pure passivity* (Husserl 1948: 110)—in the *Analysen* of 1925 not only is there established the emptiness of retention as a fundamental law of the passive genesis (Husserl 1966: 114), but it is also asserted that, unlike protension, retention is not an associative synthesis "and doesn't have in itself a directedness radiating from there toward the emptily presented past" (Husserl 1966: 119).

The reasons that move Husserl are dictated by the purpose of recognizing and demonstrating that

> That toward which the ego directs its regard—what is perceived, what is remembered, even what is retained—must already in itself be intentional, that is, must already have in its passive content a directedness toward its object (Husserl 1966: 120);

this intentional orientation could not at all be presented by retention, but only by its further intentionalization.

In BM—that represents the phenomenological–temporal background of the *Analysen*—Husserl comes to define the originary presence as a filled expectation (Husserl 2001: 6), that is not directed "only towards a new fact, but also onto the coming retentions and onto the retentions of retentions, etc." (Husserl 2001: 8). Nevertheless, retention preserves its own specificity in the terms of that differential of its passing that modifies every protension, of that differential that protension has to be able to anticipate as its own difference (Husserl 2001: 13). From the determination of the co-implication between protension and retention—which is the focus of these pages, and affects also the revision of PhZb—as well as from the new research into the *Abklangsphänomen*, it becomes possible to define the limit of consciousness as the least gradual difference between increase and extinguishing, that is, the maximum of proximity and the minimum of distance (Husserl 2001: 38–39).

(2) *Individuality*/tode ti. If already in PhZb—and in particular in the *Seefeld Manuscripts*—the description of the individual and of temporal individuality had proven central, it will be however in BM, as for that matter we have already shown, that this argument will become central, within a wider hyletic-noematic analysis. An individuality—that which can be correctly called a

"being"—is nothing other than temporal, it is the temporalization of a manner of givenness of time in a temporal manner: it is the present presenting itself, the forming of a temporal-modal givenness. That which is constituted, in fact, and that which can be constituted,

> does not possess the temporal form as something extra-essential to the being, but as something that belongs to the same and is essential to it:

that temporality is the sense of its being constituted, of its being a being (Husserl 2001: 131).

Already in *Ideas I*, the question regarding the constitution of an individual, of an object of experience, of a being in the proper sense, is split in turn into the determination of the individual essence and of the concrete essence. The last material substrates—the material not further materializable by any syntactic formation—find their place under two disjunctive main headings: "materially filled ultimated essence" and "This here!" or pure, syntactically formless, individual single particular» (Husserl 1913: 28), that is, in concrete (material) essence and individual (material) essence. Therefore, where an individual essence is filled with content, whether properly apprehended or apprehended as a spatial and temporal *plenum*, it is a concrete, a concrete individual essence, an individual (Husserl 1913: 29).

Both of them, individual essence (*the same This-here*) and the concrete essence (*just This-here*) pass through a temporal formation, but the first passes also through a further temporal-modal modalization, that of a variation of a quasi-experience. *The individual essence of an object that is each time*

> the noematic essential consistency (*noematische Wesensbestand*) is identically the same in an empirical position and in a position of quasi-experience (Husserl 2001: 290).

The individual difference of a temporal *plenum*

> is the correlate of a certain originary condition through a mode of givenness that acquires an identical correlate in the continuous change of the retentions belonging to the new "now" through each change; to the change itself there corresponds the constant alteration of the orientation, as a change of the mode of givenness of the identical (Husserl 2001: 291).

To the *Urstiftung* there corresponds the constitution of this de facto individuality, of the facticity of the *Einmaligkeit*, of the 'once presenting itself' and of the once and for all falling into continuous passing (Husserl 2001: 294). The essence of such an individuality, the *tode ti*, is the form of the singularization of the concrete, it is the de facto once having to be of each individual: it does not prejudice its whatness, nor its being-thus. It anticipates only its mode, its facticity, its contingency—understood not as the equivalent of possibility or accidentality, but only as "de facto," as modality-zero of every other modality (Bernet 2010).

Because however the phenomenological notion of metaphysics corresponds exactly to analysis, to reflection on the *Faktum*, on the de facto and on its irrationality—that is, on its impossibility of being lead back to the principle of its own reason, to the unspeakable a priority of its justification—one can well understand how the temporal treatment that Husserl dedicates to individuality cannot but be recognized as a metaphysics of individuality (Landgrebe 1972: 102–136).

(3) *Flow/Current.* Having entered into the phenomenological lexicon around the end of the first decade of the twentieth century (Bernet 1985: xxii, xlviii ff), the notion of flow has a prevailingly analogical, metaphorical meaning. It is distinguished both from *Strom*—a term taken evidently from James and attributed to the sequence constituted by lived experiences—and by *Gewühl*—with which Husserl identifies the limitless multiplicity deriving from the dissolution of apprehensions (Husserl 1973b: 84). If the *Strom* belongs to the *Erlebnis*, the *Fluss*, instead, belongs to the *Leben* (Husserl 1928: 313, 56): the first is temporal, the second quasi-temporal or intemporal. The Flow—which flows in the originary impression—

> itself is not produced; it does not arise as something produced but through *genesis spontanea*: it is primal generation [*Urzeugung*]. It does not spring [erwächst] from anything (it has no seed); it is primal creation [*Urschöpfung*] (Husserl 1928: 106).

Despite being that in which consciousness constitutes its own unity, the *Fluss* is not produced by consciousness, but is *that which has come into being alien to consciousness*, that which is felt in opposition to what is produced by the spontaneity of consciousness (Husserl 1928: 106). It therefore can present itself only reflexively, it can appear only *nach dem Konstituierten, after/in conformity with what is constituted*, remaining in its temporal non-objectivity (Husserl 1928: 79): for this reason the self-apparition of the flow does not require a further flow in which to constitute itself. The *Fluss* is the rigid and fluid reflexive form of time, the form "of a nonflowing, absolutely fixed, identical, objective time" (Husserl 1928: 67).

What happens however when this form itself becomes time and specifically, originary temporalization? The present living concrete—which already appears, in a different meaning, in BM—performs precisely this task in ST.

> The present originary has a peculiarly united living temporalization, temporal modes that flow out from the originary impression and therefore a time: the flowing impressional now and in the flowing of a One. But the flowing present is also the present of the flowing and pouring out and of the flowing into (Husserl 2006: 11–12).

This *Urgegenwart* is the originary flow of my I-am-self-temporalization, it is the reduction to the absolute *Primordium* (Husserl 2006: 118, 127). Parallel and distinct from the other hypothesis regarding temporality that is raised in ST, that is, the *monadological* one (Husserl 2006: 130–131), the primordial temporalization, causing to coincide the characteristics of the flow and those of the originary apprehension, is capable of taking up its own pre-beginning, the precedence of its being "already-always-constituted" in its continual constitution of itself (Husserl 2006: 172–173). It is in the *Primordium* that there is defined that which already in 1907 Husserl described as "the marvellous correlation between phenomenon and object of consciousness" (Husserl 1973a: 12; Husserl 1976: 184), that is, that "transcendental correlation," in the search of the origin of which the *Krisis* is initiated as if it were crossing "the threshold of the never-explored kingdom [*nie betretenen*] of the *mothers of knowledge*" (Husserl 1976: 156), of the *Mutterreich* of knowledge—a Goethian term that means the formation and transformation, *the eternal holding back of the eternal meaning*, the unlimited and ancient dominion of images."[5]

In this sense, the primordial temporality is not only a preontics—a term that exhibits a clear proof of the collaboration in this epoch between Husserl and Fink (Bruzina 2004)—but also a phenomenological archeo-mythology: the extreme attempt to elaborate a symbolism of the originary tale—or of the myth—of the origin.

5 Critiques

While in an analytical context the taking up of the phenomenology of temporality appears rather late and conditioned by the peculiar interweaving between the readings of Gurwitsch (1964) and Bergmann (1960), the critiques that instead develop within continental philosophy even precede the only publication of the results of the Husserlian research: already in 1926, in fact, Heidegger decreed in this regard a judgment of insufficiency and permanence regarding the modern metaphysical horizon from which it was never mended in the course of the following years[6] and which had a vast echo in the contemporary interpretations. If in fact in the Anglo-American thought the Husserlian analyses are, except for a few

---

[5]On the influence exercised in the 1930 s by the Husserlian readings of the works of Lévy-Bruhl, see in particular the letter of March 11, 1935 (*Briefwechsel,* Dokumente, Hua, III, 7, hrsg. von K. und E. Schuhmann, pp. 161–164), tr. in Husserl (2008). On the notion of primitive/archaic/primordial—so central in the comparison between the two authors—see the rectifications of Lévy-Bruhl himself (1949).

[6]See in this regard the letter of Heidegger to Karl Jaspers of December 26, 1926 (Heidegger and Jaspers 1990: 71), in which he recalls the occasion of the consignment, in the previous April, of the first version of *Sein un Zeit* and Husserl's contextual request for his "student" to curate the edition of the manuscripts on time. Still in 1968, in a retrospective gaze on the *Comprehension of time in the phenomenology and thought of the ontological question*, Heidegger, recalling that episode, laid claim to his decision to edit the research of the master only after his work had been given to the publisher, since they seemed to remain within the traditional concept of time, not asking how presence (*Anwesenheit*), the present, would show a characteristic of time, nor how precisely from time the sense of being would draw its determination. The Heideggerian question instead "was determined by the ontological question. It headed in a direction that would always remain extraneous to the Husserlian research into the internal consciousness of time" (Heidegger 2007: 148). Beyond the terminological and conceptual debts—from *Gegenwärtigung* to *Gewesenheit*, from *Erwärtigung* to *Zeitigung*—and from the attested Heideggerian knowledge of BM, as for that matter is declared in the *Vorbemerkung des Herausgebers* of 1928—in which there emerges the nexus between protension and retention and, what is more, the precedence of the former over the latter, which instead had long been considered as the main watershed among the analyses of the two authors—there remains intact the problem of the most profound philosophical tension of the twentieth century that cannot be listed under the title of incomprehension nor reconstructed as a mere biographical matter, or even less as a historical–political one.

exceptions (Hoy 2001), listed as a variant of the Jamesian reflections on the *specious Present* (Gallagher 1998: 32–52) or sometimes also as an integration to the Aristotelian solution adopted by Russel (1914) to resolve the aporias raised by McTaggart (1908, 1927; Broad 1923; McLure 2005: 154 ff.) and amply diffused in logical neoempiricism under the title of *Token-riflexivity* (Reichenbach 1947), in the German and French traditions in particular they are imputed—according to several variations of the Heidegerrian critical model—with a metaphysical privileging of presence (Eigier 1961).

As it is not possible to give a full account of the objections aimed at the Husserlian contributions—which cover and influence a large part of the last century of the history of philosophy—we shall try now to present their principle motives (Bernet 1985: lix–lxix).

(1) From the phenomenological impossibility to overcome the border between being and entity, demonstrated by the incapacity to pose the question regarding intentional being, there would derive the halt to the description of consciousness alone, so decisive in the analyses of temporality, in which the Cartesian inheritance would be subjected to a few hermeneutical deconstructions and, with them, the notions of idea, objectivity, monad, essence, widely diffused in the Husserlian texts (Heidegger 1979: 34–63, 148–157; Heidegger 1975: 29, 175–176).
(2) Taking up the phenomenology of time as a description of the experience of temporal spread—temporal horizon or field of presence (Merleau-Ponty 1945: 309)—we note the vicious circle according to which, in order for "a representation of sense" to be made temporal we must presuppose an a priori representation already present, thus configuring a return to Kant (Rubenstein 2001).
(3) The same blindness to history—which in no way would be remedied with the recourse in the last Husserl to the notion of *Lebenswelt*, judged rather the late result of the attempt to compare himself with *Sein und Zeit*—would be the derivative of the Husserlian ineptitude to face the essential question regarding time (Heidegger 1927: 341–369; Derrida 1962).
(4) Starting from an ontological interpretation of the nexus between time and absolute consciousness as the expression of a constant (reflexive) presence of consciousness to itself—from which this would derive the same characteristic of self-evidence—there appears inaccessible the originary passive impact of time (Levinas 1930: 56–57).
(5) Precisely by intervening on the level of intentional immanence—which does not seem damaged even by the "primordial level of lived experience of time"—it would be possible to introduce a dual correction regarding the notions of *Urimpression* and *Retention*, restoring to the former the characteristic of difference without identity, of absolute modification, of spontaneous genesis, "in which activity and passivity are completely one" (Lévinas 1974: 33), and depriving the second of the capacity to conserve and represent—objectively intact—the integrity of consciousness (Lévinas 1965: 141–145).

(6) The ambiguity of presence and absence—which still in *Retention* would be healed in favor of a primacy of the former over the latter—would be therefore lead back to the irrecuperable originary nature of deferring, from which the same ontological difference would derive and in which it would realize both a (spatial) dislocation and a (pre-temporal) delay (Derrida 1967: 53–58, 1972: 16, 158–160).
(7) That which in *Retention* would be configured as a past brought unaltered and entire into the present, would be rather recognized as the course of a trace, which could be indicated only *aprés coup* and in no way able to be lead back to its first beginning (Derrida 1972: 6–7).
(8) The taking up again of the *apeiron*—that is, of that *phänomenologisches Gewühl* spoken of above and which seems to be marginalized at least until ST and the manuscripts contemporaneous to *Krisis* (Fink 1957; Richir 1981: 183, 187)—would represent, finally, an instrument capable not only of breaking the suffocating immanent continuity of phenomenological time, but also of disposing itself, through a profound revision of Husserlian intersubjectivity, to the infinitary irruption of the Other, in which the metaphysical opening of the Ethical overcomes the identitary fixity of the Ontological (Lévinas 1961: 146, 159, 163).

The vastness of the critiques that can be inferred from this merely schematic list —involving profoundly different and hardly negotiable styles and traditions of thought—does nothing but confirm, ulteriorly and indirectly, the central role played in the last century by the laborious and not always homogeneous Husserlian analyses, so much so as to present them as a real twentieth-century encyclopedia of the philosophy of time.

References

Bergmann, G. (1960). Duration and specious present. *Philosophy of Science, 27*(1), 39–47.

Bernet, R. (1983). Die ungegenwärtige Gegenwart. Anwesenheit und Abwesenheit in Husserls Analyse des Zeitbewusstseins. Zeit und Zeitlichkeit bei Husserl und Heidegger. *Phänomenologische Forschungen, 14*, 16–57.

Bernet, R. (2010). Husserl's new Phenomenology of Time consciousness in the Bernau Manuscripts. In D. Lohmar & I. Yamaguchi (Eds.) (1–19).

Bernet, R. (1985). Einleitung in Husserl, Edmund. *Zur Phänomenologie des inneren Zeitbewusstseins. Mit den Texten aus der Erstausgabe und dem Nachlass*. Hamburg: Meiner. xv–lxxviii.

Boehm, R. (1981). Bewusstsein als Gegenwart des Vergangenen. In R. Boehm (Ed.) *Vom Geschichtspunkt der Phänomenologie. Zweiter Band: Studien zur Phänomenologie der Epoché*, 113–135. Den Haag: Nijhoff.

Broad, C. D. (1923). *Scientific Thought*. London: Routledge and Kegan.

Brough, J. B. (1989). Husserl's phenomenology of time-consciousness. In J. N. Mohanty, W. R. & McKenna (Eds.), *Husserl's phenomenology. a textbook* (pp. 249–290). Lanham (Md.): University Press of America.

Brough, J. B. (2010). Notes on the absolute time-constituting flow of consciousness. In D. Lohmar & I. Yamaguchi (Eds.), pp. 21–49.

Bruzina, R. (2004). *Edmund Husserl and Eugen Fink. Beginnings and Ends in Phenomenology 1928–1938*. New Haven: Yale University Press.

Duval, R. (1981). La durée et l'absence. Pour une autre phénoménologie de la conscience du temps. *Revue des Sciences Philosophiques et Théologiques, 65*(4), 521–572.

Derrida, J. (1962). Introduction. In E. Husserl, *L'origine de la geometrie*. Paris: PUF. English edition: Derrida, J. (1978). *E. Husserl's Origin of Geometry. An Introduction*. (P. L. John, & B. Stony Trans.). New York: N. Hays.

Derrida, J. (1967). *La Voix et le phénomène*. Paris: PUF. English edition: Derrida, J. (2011). *Voice and phenonomenon. Introduction to the problem of the sign in husserl's phenomenology* (L. Leonard, Trans.). Evanston (Ill.): Northwestern University Press.

Derrida, J. (1972). *Marges de la philosophie*. Paris: Les Éditions de Minuit. English edition: Derrida, J. (1982). *Margins of philosophy* (B. Alan, Trans.). Chicago: University of Chicago Press.

Eigier, G. (1961). *Metaphysische Voraussetzungen in Husserls Zeitanalysen*. Meisenheim am Glan: Anton Hain.

Fink, E. (1957). *Zur ontologischen Frühgeschichte von Raum-Zeit-Bewegung*. The Hague: Nijhoff.

Gallagher, S. (1998). *The Inordinance of Time*. Evanston (Ill.): Northwestern University Press.

Gurwitsch, A. (1964). *Field of Consciousness*. Pittsburgh: Dusquesne University Press.

Heidegger, M. (1979). *Prolegomena zur Geschichte des Zeitbegriffs*. GA 20. P. Jaeger. (Ed.). Frankfurt a. M.: Klostermann.

Heidegger, M. (1927). *Sein und Zeit*. Halle: Niemeyer. English edition: Heidegger, M. (1996). *Being and Time* (J. Stambaugh, Trans.) Albany: State University of New York Press.

Heidegger, M. (1975). *Die Grundprobleme der Phänomenologie* (Sommersemester 1927). GA 24. F. -W. von Herrmann (Ed.).Frankfurt a. M.: Klostermann.

Heidegger, M. (2007). *Zur Sache des Denkens*. GA 14. F. -W. von Herrmann (Ed.), Frankfurt a. M.: Klostermann.

Heidegger, M., & Jaspers, K. (1990). *Briefwechsel 1920–1963*. W. Biemel & H. Saner (Eds.). Frankfurt a. M.: Klostermann.

Held, K. (1966). *Lebendige Gegenwart. Die Frage nach der Seinsweise des transzendentalen Ich bei E. Husserl, entwickelt am Leitfaden der Zeitproblematik*. Den Haag, Nijhoff.

Hoy, R. C. (2001). Theoretical Character of Husserl's Account of Time Consciousness. In L. N. Oaklander (Ed.) (pp. 171–177).

Husserl, E. (1913). *Ideen zu einer reinen Phänomenologie und phänomenologischen Philosophie. Erstes Buch: Allgemeine Einführung in die reine Phänomenologie*. Halle: Niemeyer. K. Schuhmann (Ed.). Hua III/1 (1977). English edition: Husserl, E. 1983. *Ideas pertaining to a pure Phenomenology and to a phenomenological Philosophy*. CW II. (F. Kersten, Trans.). The Hague: Nijhoff.

Husserl, E. (1984). *Logische Untersuchungen* (1901, 1913), II, 2. U. Panzer (Ed.). HUA XIX/1. The Hague: Nijhoff.

Husserl, E. (1928). *Vorlesungen zur Phänomenologie des inneren Zeitbewusstseins*. Halle: Niemeyer. R. Boehm (Ed.). HUA X (1969). English edition: Husserl, E. (1991). *On the Phenomenology of the Consciousness of internal Time* (1893–1917) (J. B. Brough, Trans.). Dordrecht- Boston-London: Kluwer.

Husserl, E. (1976). *Die Krisis der europäischen Wissenschaften und die transzendentalen Phänomenologie*. W. Biemel (Ed.). HUA VI. The Hague: Nijhoff.

Husserl, E. (1948). *Erfahrung und Urteil. Untersuchen zur Genealogie der Logik*. Hamburg: Claassen. English edition: Husserl, E. (1979). *Experience and Judgment. Investigations in a Genealogy of Logic*, L. Landgrebe (Ed.), (J. S. Churchill & K. Ameriks, Trans.). Evanston (Ill.): Northwestern University Press.

Husserl, E. (1973a). *Idee der Phänomenologie* (1907). W. Biemel (Ed.). HUA II. , The Hague: Nijhoff.

Husserl, E. (1973b). *Ding und Raum*. Vorlesungen 1907. U. Claesges (Ed.). Hua XVI. The Hague: Nijhoff.
Husserl, E. (1966). *Analysen zur passiven Synthesis. Aus Vorlesungs- und Forschungsmanuskripten 1918–1926*. M. Fleischer (Ed.). HUA XI. The Hague: Nijhoff. English edition: Husserl, E. (2001). *Analyses concerning passive and active Synthesis. Lectures on Transcendental Logic* (1918–1926) (A. J. Steinbock, Trans.). Dordrecht-Boston-London: Kluwer.
Husserl, E. (1985). *Texte zur Phänomenologie des inneren Zeitbewusstseins*. 1893–1917. R. Bernet (Ed.). Hamburg: Meiner.
Husserl, E. (2001). *Die Bernauer Manuskripte über das Zeitbewusstsein* 1917/18. R. Bernet & D. Lohmar (Eds.). Hua XXXIII. . Dordrecht: Kluwer.
Husserl, E. (2006). *Späte Texte über Zeitkonstitution* (1929–34). Die C- Manuskripte. D. Lohmar (Ed.). HUA Materialen VIII. Dordrecht: Springer, 2006.
Husserl, E. (2008). Letter to Lucien Lévy-Bruhl. (L. Steinacher & D. Moran, Trans.). *The New Yearbook for Phenomenology and phenomenological Philosophy 8*, 1–6.
Ingarden, R. (1922). Intuition und Intellekt bei H. Bergson. *Jahrbuch für Philosophie und phänomenologische Forschung, 5*, 286–461.
James, W. (1890). *Principles of Psychology*. New York: Holt & Co.
Kortooms, T. (2002). *Phenomenology of Time: E. Husserl's Analyses of Time-Consciousness*. Dordrecht: Springer.
Landgrebe, L. (1972). *Faktizität und Individuation. Studien zu den Grundfragen der Phänomenologie*. Meiner: Hamburg.
Lévinas, E. (1930). *La théorie de l'intuition dans la phénoménologie de Husserl*. Paris: Alcan.
Lévinas, E (1965). Intentionnalité et sensation. *Revue internationale de philosophie* 71–72: 34–54. English edition: Lévinas, E. (1998). Intentionality and Sensation. In *Discovering Existence with Husserl* (R. A. Cohen & M. B. Smith, Trans.), 135–150. Evanston (Ill.): Northwestern University Press.
Lévinas, E. (1961). *Totalité et infini. Essai sur l'extériorité*. La Haye: M. Nijhoff. English edition: Lévinas, E. (1969). *Totality and infinity. An essay on exteriority* (A. Lingis, Trans.). Pittsburgh: Duquesne University Press.
Lévinas, E. (1974). *Autrement qu'être ou au-delà de l'essence.*. La Haye: M. Nijhoff. English edition: Lévinas, E (1981). *Otherwise Than Being or Beyond Essence* (A. Lingis, Trans.). Dordrecht-Boston-London: Kluwer.
Lévy-Bruhl, L. (1949). *Carnets* (1938–39). Paris: Puf.
Lohmar, D. (1996). Zu der Enstehung und den Ausgangsmaterialen von E. Husserls Werk "Erfahrung und Urteil". *Husserl Studies, 13*(1), 31–71.
Lohmar, D., & Yamaguchi, I. (2010). (Eds.) *On Time. New Contributions on the Husserlian Phenomenology of Time*. Dordrecht-Heidelberg-London-New York: Springer.
McLure, R. (2005). *The Philosophy of Time. Time before Times*. London-New York: Routledge.
McTaggart, J. M. E. (1908). The Unreality of Time. *Mind, 17*, 457–474.
McTaggart, J. M. E. (1921–1927). *The Nature of Existence* (Vol. 2). Cambridge: Cambridge University Press.
Meinong, A. (1899). Über Gegenstände höherer Ordnung und deren Verhältnis zur inneren Wahrnehmung. *Zeitschrift für Psychologie und Physiologie der Sinnesorgane, 21*, 182–272.
Mensch, J. (2010). Retention and Schema. In D. Lohmar & I. Yamaguchi, pp. 153–168.
Merleau-Ponty, M. (1945). *La Phénoménologie de la perception*. Paris: Gallimard. English edition: Merleau-Ponty, M. (2002). *Phenomenology of Perception*. (C. Smith, Trans.) London-New York: Routledge.
Oaklander, L. N. (Ed.). (2001). *The Importance of Time*. Dordrecht-London: Kluwer.
Reichenbach, H. (1947). *Elements of symbolic logic*. New York: Free Press.
Richir, M. (1981). *Recherches phénoménologiques I, II, III. Fondation pour la phénoménologie transcendentale*. Ousia. Bruxelles.
Rubenstein, E. M. (2001). Experiencing the future: Kantian thoughts on Husserl. In Oaklander L. N. (Ed.), pp. 159–169.

Russell, B. (1914). On the Experience of Time. *Monist, 25*, 212–233.
Schnell, A. (2004). *Temps et phénoméne. La phénoménologie husserlienne du temps* (1893–1918). Hildesheim: Olms.
Sokolowski, R. (1974). *Husserlian meditations. how words present things*. Evanston (Ill.): Northwestern University Press.
Stern, W. (1897). Psychische Präsenzzeit. *Zeitschrift für Psychologie und Physiologie der Sinnesorgane, 13*, 325–349.
Zahavi, D. (2011). Objects and levels. reflections on the relation between time-consciousness and self-consciousness. *Husserl Studies, 27*, 13–25.

L'évasion de l'être. Jean-Paul Sartre and the Phenomenology of Temporality

Armando Mascolo

> *Souviens-toi* que le Temps est un joueur avide qui gagne sans tricher, à tout coup! C'est la loi. Le jour décroît; la nuit augmente; *Souviens-toi!* Le gouffre a toujours soif; la clepsydre se vide.
>
> Charles Baudelaire, *L'Horloge*.

Abstract Sartre fits fully within the phenomenological tradition inaugurated by Husserl, although he somewhat reelaborates it in an original way, on the basis of Heidegger's philosophy, with the aim of outlining, in a first stage of his thoughts dating back to the publication of Being and Nothingness (1943), the features stemming from his peculiar atheistic existentialism. Subsequently, in the mature stage of his intellectual itinerary, Sartre will attempt to combine the existentialist ideas with the basic principles of Marxism, a synthesis that will create important works such as Search for a Method (1957) and Critique of Dialectical Reason (1960). This chapter analyses time from the phenomenological perspective of the Sartrean ontology of temporality. This analysis allows to conceive the typically human "existential time" as a permanent existence out of oneself. For this purpose, we will retrace the fundamentals of Sartre's phenomenological ontology outlined in his most important work of 1943, Being and Nothingness, focusing in particular on the structures of consciousness understood as "être-pour-soi" ("being-foritself"), i.e. the conditions of man's "absolute freedom", which Sartre refers to when he shows the one "pour-soi" as the being who is pure nothingness.

The philosophical career of Jean-Paul Sartre (1905–1980) can be rightfully identified with the phenomenological tradition inaugurated by Husserl, which he revised in a rather personal way in the light of the philosophy of Heidegger. Sartre shows a clear desire to outline the core features of his own particular vision of atheistic existentialism in an initial phase of his thought which dates back to the publication of his philosophical masterpiece, *Being and Nothingness* (1943). Afterwards, starting with the mature phase of his speculative journey, the so-called phase of the *intellectuel engagé*, Sartre focused all his energies on an attempt to combine his

A. Mascolo (✉)
Institute for the History of Philosophy and Science in Modern Age, National Research Council, Via Porta di Massa 1, 80133 Naples, Italy
e-mail: armandomascolo@hotmail.com

F. Santoianni (ed.), *The Concept of Time in Early Twentieth-Century Philosophy*, Studies in Applied Philosophy, Epistemology and Rational Ethics 24,
DOI 10.1007/978-3-319-24895-0_8

existentialist ideas with the fundamental principles of Marxism, in a completely original approach that would lead to important works such as *Search for a Method* (1957) and *Critique of Dialectical Reason* (1960).

An eclectic and untiring writer, Sartre became the author of a complex and protean endeavour, the expression of an inexhaustible intellectual vivacity that drove him to stubbornly confront the most disparate theoretical questions and try his hand at various literary genres, albeit with varying results, ranging from romance to theatrical drama, and from purely philosophical essays to journalistic articles.

Within this broad scenario, a unique and significant space was given to the problem of temporality, a problem that Sartre felt the need to confront with a certain urgency while serving in the armed forces during the Second World War. He openly admits this in his *War Diaries*, written between November 1939 and March 1940, where, following the example of the *Journals* of Andrè Gide, the philosopher not only faithfully noted down episodes he witnessed during his daily military duties, but also extracts from his personal correspondence and philosophical reflections on the meaning of war, human existence and history: "I feel strangely bashful about embarking on a study of temporality. Time has always struck me as a philosophical headache, and I'd inadvertently gone in for a philosophy of the instant (which Koyré reproached me for one evening in June 1939)—as a result of not understanding duration. [...] And, behold, I now glimpse a theory of time! I feel intimidated before expounding it, I feel like a kid" (Sartre 1984: 256–257, 208–209). From this perspective, the *War Diaries* can be considered an important moment of transition between the young Sartre, who entrusted his reflections to the literary character Antoine Roquentin, the hero of his first novel *Nausea* from 1938, and the now mature existentialist philosopher, who presents and discussed his theories through the dense pages of *Being and Nothingness*.

Actually, the question of temporality was already present in the narrative dimension of *Nausea*, (Cf. Sartre 2000), where Sartre's reflection seems to be guided mainly by thoughts concerning temporality and the temporal nature of freedom (cf. Clayton 2009). The philosophical topic of the "lived moment", as a paradoxical contact between time and eternity, assumes a central importance in the story, where the instant is embraced and valued by Sartre as the fundamental *moment* of existence, and is therefore viewed as being radically different from the "mathematical instant", considered to be the *limit* to the divisibility of quantitative time (cf. Pieri 1998: 27–30).

It was only with *Being and Nothingness*, however, that the topic of time was given a systematic and in-depth examination (cf. Bukala 1975). In the essay on phenomenological ontology, Sartre made a lucid, meticulous and penetrating investigation on the question of temporality, contained in a dense chapter written in the light of the ontological categories developed in the first part of the work, regarding the structures of consciousness or "being-for-itself". It is therefore necessary to start from these later works in order to fully understand the significance of the phenomenological perspective of Sartre's ontology of temporality, as "most features of the for-itself involve temporalization" (Manser 1989: 25).

The theoretical framework of Sartre's ontology is based on a Cartesian dualism, according to which being is divided into two "regions" that are absolutely distinct and irreducible, even though both are characterized on a metaphysical level by contingency and the absurd: on the one hand, there is the being of the phenomenon or "being-in-itself" (*être en soi)*, i.e., the being of things that appear to human consciousness and that cannot be reduced to their simple appearance without the risk of reviving idealism; and, on the other hand, there is the "being-for-itself" (*être pour soi*), or the being of the consciousness to which things appear and which, in turn, cannot be reduced to the world that appears to it, with the danger, in this case, of ending up in realism. The basic motive behind Sartre's book is therefore to return to being in order to describe its structures, with the precise intention of outlining an ontology that can relate the two distinct regions into which being appears to be divided from the origins. From this perspective, Sartre's approach can be seen as an ontology of reality not in its mere givenness, but as it appears in its totality, as a phenomenon classified, described and named as being. In other words, it is a theory of "appearance" as being, or of being within the limits of appearance. Sartre, in fact, believed that the most significant progress made by modern thought, to which he himself refers, was to have reduced "the existent to the series of appearances which manifest it" (Sartre 1993: 3). It thus follows that "the dualism of being and appearance is no longer entitled to any legal status within philosophy". Having said this, Sartre resolutely emphasizes that he no longer believes "in the being-behind-the-appearance", inasmuch as "the being of an existent is exactly what it *appears*. […] What it is—concludes the French philosopher—it is absolutely, for it reveals itself as it is. The phenomenon can be studied and described as such, for it is *absolutely indicative of itself*" (Sartre 1993: 4). This also explains the subtitle of Sartre's work, *A Phenomenological Essay on Ontology*, since ontology is interpreted in a phenomenological sense, namely as a description of structures determined and developed by thought on the basis of experience, or, as a "description of the phenomenon of being as it manifests itself; that is, without intermediary" (Sartre 1993: 7).

The two different "aspects of being" examined by Sartre have diametrically opposed ontological characteristics and are therefore irreconcilable. Sartre provides a precise and philosophically detailed description of them, perfectly in line with the phenomenological framework within which his reflection is made.

Being-in-itself, the French philosopher states, is what it is in its naked and brutal immediacy, i.e. it appears infinitely compact, dense, filled with itself and therefore opaque, solid, immobile, without relations, characterized by gratuitousness and timelessness. Thus it is absolute positivity and identity, in that it cannot be other than what it already is. Sartre provides a very powerful and theoretically effective image of being-in-itself that is worth quoting in its entirety in order to fully capture its various nuances: "Being is not a connection with itself. It is *itself*. It is an immanence which cannot realize itself, an affirmation which cannot affirm itself, an activity which cannot act, because it is glued to itself. […] Being is opaque to itself precisely because it is filled with itself. This can be better expressed by saying that *being is what is it*. […] Being-in-itself has no *within* which is opposed to a *without* and which is analogous to a judgment, a law, a consciousness of itself. The in-itself has nothing secret; it is *solid* (*massif*). […] Being is isolated in its being and … does not enter into any connection with what is not itself. […] It is full positivity. It knows no *otherness* […] It is itself indefinitely and it exhausts itself in being. From this point of view we shall see later that it is not subject to temporality. […] Being can neither be derived from the possible nor reduced to the necessary. […] Being-in-itself is never either possible or impossible. It *is*. […] Being is. Being is in-itself. Being is what it is. These are the three characteristics which the preliminary examination of the phenomena of being allows us to assign to the being of phenomena" (Sartre 1993: 27–29). This description, set in a radically anti-idealist perspective, is intended to highlight the complete self-sufficiency of being-in-itself in respect to consciousness and is also interesting in that it is an expression of the intensely atheistic and anti-creationist position that characterizes Sartre's work. If being-in-itself appears as self-sufficient in relation to consciousness, and thus humanity is entirely inessential to it, the same fate also befalls God, in that being is uncreated, and therefore not put in place by any God and not even by itself. Being-in-itself, in ultimate analysis, is neither a creature nor self-created; it is pure "self-identity", and as such it is opacity, fullness, lack of all otherness, irreducible to the possible or the necessary. In other words, it is absolute contingence, without any reason for being, without origin, without destiny and without end: it simply is.

No matter how self-sufficient it may be with regard to consciousness, the being of the phenomenon nevertheless requires someone to appear to: the *percipi* requires the *percipiens*, i.e. it requires a consciousness. Referring to Husserl, Sartre affirms that this, for its part, is always "intentional", i.e. it is always conscious of something that is not consciousness, in that it is always oriented, projected towards what is other than itself, towards the in-itself, that is to say, towards the being of things, towards the being of phenomena that appear to consciousness. In other words: I am aware of the objects of the world, but none of these objects is my awareness. Thus, a particular dialectic is established between the two dimensions of being, as two entities which, although opposites, are in strict relation, a relation that arises from consciousness itself, in that it is by essence aware of something and is relation with the being (cf. Schnaith 1970: 101–102).

Consciousness is distinguished not only through intentionality, but also by its presence to itself, by its being a pre-reflective consciousness of self. This presence to itself implies a sort of rupture, doubling or interior separation within the being of consciousness. The latter, in fact, never manages to coincide with itself, in that it is a continuous negation of itself, a transcendence of its own being already. For this reason, consciousness qualifies as an essentially incomplete being, continually searching for completion and never satisfied. Therefore, Sartre sees human reality as a radical "lack of being" (*manque d'être*), an unceasing tension towards being, or a "desire of being" (*désir d'être*). Moreover, the fact that human reality is substantially a lack is shown by the existence of desire as a human fact (cf. Kremer Marietti 2005). With regard to this, Sartre writes: "If desire is to be able to be desire to itself it must necessarily be itself transcendence; that is, it must by nature be an escape from itself toward the desired object. In other word, it must be a lack, but not an object-lack, a lack undergone, created by the surpassing which it is not; it must be its own lack of. Desire is a lack of being. It is haunted in its inmost being by the being of which it is desire. Thus it bears witness to the existence of lack in the being of human reality" (Sartre 1993: 137).

The human separation of self from self is, in Sartre's own words, a continuous self-annihilation; it is nothingness as the nullifying power of consciousness. Nothingness, in ultimate analysis, is the necessary condition of the for-itself, i.e. of human consciousness, which is essentially seen as "nihilation" (*néantisation*) of the in-itself. As "presence to things", consciousness has the power to give meanings to them, despite them being essentially gratuitous and meaningless, or to transcend them, to "annul them" as pure data and affirm itself as having sole responsibility for their meaning. In order to do this, however, consciousness must be in a particular condition, as it must be absolutely free. By freedom, Sartre means precisely that possibility of nullification of the self and the world that is the very structure of human existence. Consciousness, ultimately, transcends the being and reveals itself essentially as project and freedom, a thrust towards possibilities, in an ultimately vain search for a foundation and absolute meaning for itself and the world.

It is in this ontology of consciousness and freedom that Sartre introduces the topic of time. His basic theory is that temporality comes to being thanks to being-for-itself. From this perspective, it has been rightly pointed out that "freedom, choice, nihilation, temporalization are all one and the same thing" and therefore "the chapter on temporality is the hinge on which the argument of *L'Être et le néant* pivots" (Manser 1989: 25). Time, in Sartre's perspective, draws its origin from the nullifying function of the for-itself, indeed, it represents one of the fundamental ways by which it is differentiated from the in-itself. Only a being that is internal negation, transcendence and project can in fact be temporal in the original sense. This is what consciousness is, for which it only makes sense to speak of present, past and future. Temporality, therefore, represents the very essence of consciousness. Time does not exist by itself but only as a process of temporalization of consciousness, which, in its constant tension, continually refers to being, which it is not, thereby affirming itself as nothing.

In his investigation of the problem of time, Sartre begins from a preliminary phenomenological description of the individual temporal dimensions (past, present and future), in the awareness that each of these, in no sense independent from the others, must be viewed against the background of temporal totality (cf. Quaglia 1980: 51–52).

The main characteristic of the for-itself, as we have shown previously, is that of not being what it is. Due to this condition, the for-itself relates to the before, with the *ekstasis* of the past. According to Sartre, the mistake constantly repeated over the course of the history of ideas has always been that of considering the past in itself, as remote from the present to which it refers. In reality, states the French philosopher, the past is never something autonomous and independent, but is always the past of this present. The past, moreover, is always my past; a past in itself, or an abstract and universal past, does not exist: "There is not first a universal past which would later be particularized in concrete pasts. On the contrary, it is *particular pasts* which we discover first" (Sartre 1993: 165). Therefore, in order to fully capture the sense of the past, and the meaning of time in general, we have to start from the consciousness of the individual. For this reason, a consciousness cannot be said to "have" a past; "to have a past", in fact, implies possession, or a merely extrinsic relationship: "One can not 'have' a past as one 'has' an automobile or a racing stable. That is, the past can not be possessed by a present being which remains strictly external to it as I remain, for example, external to my fountain pen. In short, in the sense that possession ordinarily expresses an external relation of the possessor to the possessed, the expression of possession is inadequate. External relations would hide an impassable abyss between a past and a present which would then be two factual givens without real communication" (Sartre 1993: 166). The past, however, is the past of the present of a consciousness; it is my past, the past of a present that I am. Basically, I am my past, and in this sense I live it, I carry it within me as an integral part of my person, assuming full responsibility for it. From another perspective, however, I am not my past, in that I was it and I adopt a stance of negation and rejection towards it: it is there, distant from my present and other than what I am at present. The past, therefore, on the one hand, is something that is in me, which I cannot avoid and thus have to make mine; on the other, it is an in-itself that is behind me, incapable of making any actual impact on my present. In conclusion, if, on the one hand, I am my past, and my contingency and the "facticity" of my existence consist in this, on the other, I continually transcend this "facticity" by introducing an unbridgeable "crack" between what I was and what I am (cf. Sartre 1993: 173-175).

With regard to the future, just like the past, it is a for-itself that incarnates the features of the in-itself: it is that dimension of the for-itself by reason of which it is its own future, towards which it transcends itself, grasping itself as an essential lack that has to be filled. The future, in other words, is that which the for-itself lacks in order to be fully itself; it is that infinite horizon of possibilities of the for-self that are destined to only be partially fulfilled. "I project myself—Sartre says—toward the Future in order to merge there with that which I lack; that is, with that which if synthetically added to my Present would make me be what I am" (Sartre 1993: 168). The for-itself, however, is condemned to never completely be its future, since its freedom implies the constant possibility of following new paths and abandoning certain projects in order to create new ones: "The future does not allow itself to be rejoined; it slides into the Past as a bygone future, and the Present For-itself in all its facticity is revealed as the foundation of its own nothingness and once again as the lack of a new future. Hence comes that ontological disillusion which awaits the For-itself at each emergence into the future" (Sartre 1993: 169).

The present, according to Sartre, has been traditionally made to coincide with being, given that the past is no longer and the future is not yet. It is clear, however, that if the present is considered in this way, detached from the other two temporal dimensions, it vanishes in an instant, which is a mere abstraction. The present, however, shares the same ontological structure as the for-itself and consists in the absolute presence of the subject to other beings, to being-in-itself. However, being present to the in-itself, and continuously transcending towards it, implies that duplication or fragmentation of the for-itself that is manifested in the form of negation, in the *évasion de l'être*: "The present is precisely this negation of being, this *escape from being* inasmuch as being is *there* as that from which one escapes. The for-itself is present to being in the form of flight; the Present is a perpetual flight in the face of being" (Sartre 1993:179). Contrary to what common sense would have us believe, the present—strictly speaking—does not exist, but is manifested as a continuous "escape" from the self.

Consciousness as nullification and dispersion, however, is not only the source from which the three temporal *ekstases* arise, but also ensures their unity. The phenomenological analysis of the individual temporal dimensions, in fact, is just a "provisional" step, for the sake of their ontological unity within a single process of temporalization. The *ekstatic* unity of the latter constitutes the very essence of the for-itself, of that being which is always beyond itself, and, by its very nature, "diasporic" or, in other words, dispersed in its temporal *ekstasis*, despite being the unity and foundation of the dispersion (cf. Quaglia 1980: 60–67).

Although a dominant dimension does not exist within the single process of temporalization, such as to have an ontological priority over the others, from a phenomenological perspective, Sartre, unlike Heidegger, shifts the centre of the temporal dynamic from the future to the present (cf. Tortolone 1993: 100–101). The present is in fact emblematic of the very essence of consciousness as dispersion and nullification, and is therefore indispensable to understand temporality in its total and unified form. Despite the importance given by Sartre to temporality as the internal structure of consciousness, in reality, towards the end of his analysis, the image emerges of a for-itself constantly striving to negate the time lived as its own real condemnation (cf. Moravia 2010: 52), in the ultimately vain attempt to place itself outside of it.

References

Bukala, R. C. (1975). Sartre's existentialist view of space and time. *Philosophical Studies, 24*, 166–180.

Clayton, C. (2009). Nausea, Melancholy and the internal negation of the past. *Sartre Studies International, 15*(2), 1–16.

Kremer Marietti, A. (2005). Jean-Paul Sartre et le désir d'être. Une lecture de L'Être et le néant. L'Harmattan: Paris.

Manser, A. (1989). Sartre on temporality. *Journal of the British Society for Phenomenology, 20*(1), 23–32.

Moravia, S. (2010). *Introduzione a Sartre*. Laterza: Bari-Roma.

Pieri, S. (1998). Tempo, istante, eternità nel pensiero di Sartre. *Giornale di Metafisica, 10*, 11–30.

Quaglia, P. (1980). La concezione del tempo nella filosofia di Sartre. *Filosofia, 31*, 52–82.

Sartre, J. -P. (1984). *War diaries: Notebooks from a Phoney War (November 1939–March 1940)*. (H. Quintin, Trans.). London: Verso.

Sartre, J. -P. (1993). *Being and nothingness. A phenomenological essay on ontology*. (H. E. Barnes, Trans.). New York, Washington: Square Press.
Sartre, J. -P. (2000). *Nausea*. (R. Baldick, Trans.). London: Penguin Books.
Schnaith, N. (1970). Tiempo y subjetividad en Sartre. *Cuadernos de Filosofía, 10*, 90–111.
Tortolone, G. M. (1993). *Invito al pensiero di Sartre*. Milano: Mursia.

The Time of the Body in Maurice Merleau-Ponty

Maria Teresa Catena

Nothing lasts and yet nothing passes either,
and nothing passes just because nothing lasts
Philip Roth, *The Human Stain*

Abstract The starting point of Merleau-Ponty's reflection on time is the notion of functioning intentionality observed in its specific application as a perceptive activity. Through an original treatment of the notion of the perceptual field, the French philosopher describes the activity that, within this field, a particular protagonist carries out, namely one's own body: a particular kind of extension thanks to which it is possible to overcome all those dualistic prejudices that abstractly contrast the subject, or consciousness, with the world and its objects. Instead, in the perceptual field described in the pages of *Phenomenology of Perception*, the body and the world are "born" in unison. They are reciprocally constituted, to such a point that we can no longer speak of a pure subject or of a disembodied consciousness, separated from objects, but must speak of a corporeal knowledge that is always in relationship with a world that, for its part, finds its objective dimension only by abstracting from that original conferring of meaning attributed to it by bodily action. Now, it is on the basis of this that the analysis of temporality is carried out. The present is a nexus of time that one's own body lives and exists in the perceptual field. More than a representation, time, with its dimensions, is a concrete thickness that is stratified a-thematically in the activity of the body that always inheres in the world. It thus involves not a linear becoming, a summation of instants, but a flow, a continual transition that, from the present, allows access to the past and future which in turn emerge as stratified in the lived time of the present.

Perhaps the best way to introduce the prospect that Merleau-Ponty has on time, is to ask two simple questions, similar yet inverse to each other.

The first reads roughly like this: would there be time, if we were not there? Would it flow like the river of Heraclitus, as a flowing substance, if there was no observer able to verify its passage?

M.T. Catena (✉)
Philosophy Section, Department of Humanities, University of Naples Federico II,
via Porta di Massa 1, 80133 Naples, Italy
e-mail: mt-catena@libero.it

F. Santoianni (ed.), *The Concept of Time in Early Twentieth-Century Philosophy*,
Studies in Applied Philosophy, Epistemology and Rational Ethics 24,
DOI 10.1007/978-3-319-24895-0_9

Similarly, and vice versa: if there was no world, would there be time?

Well, the well-known Merleau-Ponty, answering the first question, affirms: “If I consider the world itself, there is simply one indivisible and changeless being in it,” a simple series of nows, the eternal survival of this “too much of a plenum,” with no past and no future (Merleau-Ponty 1945: 477 and 478).

The time, in fact, always involves “a view of time.”

Thus, given that “change presupposes a certain position which I take up and from which I see things in procession before me,” there is no time and no event “without someone to whom they happen and whose finite perspective is the basis of their individuality” (Merleau-Ponty 1945: 477).

Given these assumptions, we must guard against identifying subjectivity as the center of time, to the point of making it an exclusive matter of our consciousness.

Time, in fact, we read in the *Phenomenology of Perception*, cannot be understood apart from physiological theories which trace memory back to brain traces and to other corporeal devices.

But not only that.

Even more misleading and erroneous of this tracing back to organic factors would be the reduction of memory to the conservation of a psychic trace. Making such a statement means in fact thinking of the dimensions of the past and future as simple concepts, obtained by abstraction starting from our individual perceptions and our individual memories; it means, in other words, to make of time a given *of* consciousness. However, continues the philosopher, to be able to perform this operation, “the subject must *in order to be able* to be present in intention to the past as to the future”: “must not be himself situated in it” (Merleau-Ponty 1945: 481). In this way, however, time, as a given *of* consciousness, ends up being given *by* consciousness. Rather than thinking of it, in short, the timeless subjectivity becomes what places it, constitutes it.

As paradoxical as it may sound, there happens here something similar to what we noted previously: it is the very phenomenon of time, with its different dimensions, that is destroyed. Before this constituting consciousness, contemporary to all times, we would in fact find ourselves in front of a kind of eternal present: “time as the immanent object of a consciousness is time brought down to one uniform level, in other words it is no longer time at all” (Merleau-Ponty 1945: 481 and 482).

Similarly, we must therefore respond negatively to the second question and say, in a word: no, nothing could be said of time if there was not a world there, outside of us, that flows before our eyes.

So, if there is no absolute or objective time, the ultimate guarantee in which our subjectivity finds sense and meaning, it is not even possible to speak of a time that is established in a constituent activity described as independent of any relationship with the world.

And so?

What can we say about time?

Whence the wellspring of its flow?

Now, if there is an indication that the thought of Merleau-Ponty intends to give, if there is an effort that his philosophical journey moves us to take, it is to bypass every prospect that imposes its reasoning on the juxtaposition between an a priori and an a posteriori. As a good phenomenologist, instead, he suggests starting with the *relation*, showing thus that he agrees with what Husserl had written about the intentionality of the *cogito*: "in general, there belongs to the essence of each current *cogito* to be aware 'of' something" (Husserl 1913: 76).

It is thus by moving in this direction that we must come to understand the issue in question. In other words: in order to grasp the way in which Merleau-Ponty thinks of time we must first start with taking off the clothes that Western metaphysics has sewn on, and through the method of *epochè*, put "in abeyance the assertions arising out of the natural attitude, the better to understand them" (Merleau-Ponty 1945: VII).

By doing so we can then observe that the relation of the subject and the world is not the product of two terms that are added, different from each other: on the contrary, the philosopher notes, we must bring out that "the relation between subject and world (…) are strictly bilateral" and that, therefore, "the unity of consciousness (…) is achieved simultaneously with that of the world" (Merleau-Ponty 1945: X). In short, neither the subject nor the world belong to themselves: there is, instead, a being given together, a contemporaneous emergence of the subject and the world.

Of course, the first consequence of this way of seeing things will be to rethink the whole idea of subjectivity and, correspondingly, of the world. More than being "the basis of the relatedness" or "an impregnable subjectivity, as yet untouched by being and time" (Merleau-Ponty 1945: X–XI), we "are through and through compounded of relationships with the word" (Merleau-Ponty 1945: XIV). Similarly, "the word is not an object such that I have in my possession the law it's making; it is the natural setting of, and field for, all my thoughts and all my explicit perceptions" (Merleau-Ponty 1945: XI–XII).

Now, there is no doubt that perception is the intentional act that among all of them gives us back our radical *in der-Welt-Sein* (Heidegger 1927). In a revival of the Husserlian notion of acting intentionality, Merleau-Ponty affirms that "perception" must be "defined as access to truth" (Merleau-Ponty 1945: XVIII).

We have therefore come to identify the dimension that it is necessary to cross to find the threads that lead us to the conception of temporality: "not from a central I" nor from the outside world, but it is from perception that "there depart these protentions and retentions that allow me to count on a future and to maintain a tradition of the past" (Brena 1969: 128).

To say perception, for Merleau-Ponty, means to say many things.

Since the time of *The Structure of Behavior* (Merleau-Ponty 1942), it is brought back to the *perceptual field* or *phenomenal field*.

Recovering from *Gestalt psychology* (Koffka 1935; Köhler 1929) a series of concepts, Merleau-Ponty highlights and questions the prejudice of sensation. There do not exist, except as abstractions, neither the pure *what* of a-spatial and a-temporal sensation nor the immediacy with which it would be imprinted on us: it is rather the structure, the layout, the articulation of the parts of the object that are perceived. It is the whole and the set, not the single impression, to constitute in the first instance the perceptual event.

However, on the "laws of a descriptive nature that show the forms of ordering of perceptual materials Merleau-Ponty does not dwell except incidentally. It is no coincidence: for Merleau-Ponty, outlining a phenomenology of perception does not mean showing the structures that make possible the experience, but drawing attention to the fact that the blind dimension of the stimulus is still passed over in the direction of a *sense*, that arises because the *subjectivity* recovers and amplifies what is contained in the perceptive scene" (Spinicci 2000: 169).

Central therefore, in this field of perception, is subjectivity; indeed, it would be better to say that particular type of observer who is their *own body*.

But what does Merleau-Ponty mean when he speaks of his own body?

Certainly a very particular kind of extension. If in fact the object is that invariable structure that is always constituted through the changing of perspectives, the body is "an object which does not leave me," whose permanence is a permanence on my side. In other words: if the object is something that is in front of me, and it is such only if it can be moved away and found in my field of vision, the body, on the contrary, is always present for me, always on my side: "to say it is always near me, always there for me, is to say that it is never really in front of me, that I cannot array it before my eyes, that it remains marginal to all my perceptions, that it is with me" (Merleau-Ponty 1945: 103 and 104).

This peculiarity is also evident from another character of my own body: its being, that is, an undivided unity. After all, in fact, I never perceive my body as an 'aggregate of juxtaposed organs' or a sum of parts, but as a total body, a system, a synesthetic and inter-sensory unit; a corporeal schema, to use a term much in vogue at the time (Schilder 1935).

So, if it is a question of extension, it is a very special extension: we could call it the *zero-point* or *point of resistance* from which the possible perspectives are given on the objects and their very permanence.

Of course, this is not a fixed point but rather a *mobile-point*.

Repeatedly Merleau-Ponty refers to the motor intentionality of his own body and emphasizes how its form and structure have an essentially dynamic character. It goes without saying that, given its features, its movements will never be comparable to those of a piece of furniture in physical space: my body is "a system of possible actions, a virtual body with its phenomenal 'place' defined by its task and situation. My body is wherever there is something to be done" (Merleau-Ponty 1945: 291).

Such doing is not, evidently, for the philosopher, to be understood as the mechanical movement of a part of the body, nor as a gesture directed by a representation. Already in the simple act of a tailor who, for example, in sewing a fabric takes up a pair of scissors, there is something much more: his hands, in fact, are not simple anatomical parts of a whole, but 'power', already 'mobilized' for the act; in the same way in which the subject, to move them, has no need to represent them.

The motility of my own body is just an example and index of a subjectivity thought beyond the classic definitions; beyond, that is, any view that reduces the body to a mere *res extensa* and conceives the representative activity in the disembodied terms of a consciousness pure and transparent to itself. Instead, with its doing, which is always concrete living in the world, the body shows and leads back to a non-conscientious capacity to give sense and meaning: meaning and significance that—mind you—are not so much an accessory as the source mode with which the world is presented to us, as existential space and environment for behavior.

A new form of constitution, quite different from that implemented by the transcendental consciousness, is described in these pages.

For if there is one point on which Merleau-Ponty beats firmly it is the co-originality of the body and the world in the field of perception.

Giving a sense, in fact, does not mean to create the world, to place it as one's own act: it involves rather a grasping and taking up an existential, symbolic and historic order that we ourselves do not create.

And then, of course, perception teaches us that "to understand is ultimately always to construct, to constitute, to bring about here and now the synthesis of the object"; and also teaches us to operate "a universal setting in relation to the world." However, it is thanks to this that we can find out about "the subject himself (…) the presence of the world" and, finally, to stop thinking of it as "a synthetic activity" and begin to consider it one's own body, "*ek-stase,*" to be "orientated or polarized in the direction of what he is not."

Here then is what we understand from the observation of the perceptual field: "the world is inseparable from the subject, but from a subject which is nothing but a project of the world"; in the same way in which "the subject is inseparable from the world, but from a world which the subject itself projects" (Merleau-Ponty 1945: 499 and 500).

Given these premises, the moment has come to return to time.

There is no doubt, in fact, that it is only from this description of the perceptual field, as a place of constitutive inherence between the self and the world, that we can understand the perspective of Merleau-Ponty on time.

In the first instance, evidently, these aforementioned assumptions lead us to understand that the node of time is the present, understood as the current perceptual field.

"It is in my 'field of presence' in the widest sense—this moment that I spend working—(…) that I make contact with time, and learn to know its course." And again: "Everything, therefore, causes me to revert to the field of presence as the primary experience in which time and its dimensions make their appearance unalloyed, with no intervening distance and with absolute self-evidence" (Merleau-Ponty 1945: 483).

It is clear, however, that this does not mean to crush the being to the world onto the present. On the contrary, the now is the moment of the recovery of the horizons of past and future, always underlying it: "it is here that we see a future sliding into the present and on into the past" (Merleau-Ponty 1945: 483).

It is equally clear that Merleau-Ponty does not read the dimensionality of time in a representative sense.

The three dimensions, notes the philosopher, are not given me in virtue of discrete acts; in other words, I do not *represent* my day, in the same way that I do not *think* of the evening that will come, nor do I *place* in any way the present: "I do not pass through a series of instances of now, the images of which I preserve and which, placed end to end, make a line" (Merleau-Ponty 1945: 484). In short, in the perceptual field in which I am, I do not perform an intellectual synthesis, I do not go back mentally from the present moment to the past or from the present to the future.

Mind you: this does not mean that time does not require the synthesis of identification. It can certainly happen in fact that I hesitate on the date of a memory or that I have in mind a certain scene and do not know to which point of time to connect it. Well, in these cases it is always possible to position these memories through an intellectual 'going-back' able to reconstruct the causal order of the events.

Evidently, however, this has nothing to do with grasping "the concrete origin of the memory" (Merleau-Ponty 1945: 485).

Here, the movement is very different.

In this case, in fact, my present passes over itself and goes to a past not trying to think of it, but *touching* it where it is. More than a memory expressed, there occurs here a non-thematic reaching of lost time; not a reactivation of the *memory* of fear and hope, but a reactivation of the *same current* of fear and hope. It is making the past appear '*in person*', it is *feeling* it, rather than thinking of it. In short, I do not remember my workday by focusing mentally on what I did, recalling the significant episodes of it; sure, I can do it, but these thematic memories, these intentionalities of act, find their wellspring in an acting intentionality, that would be to say, in feeling, for example, the weight of the day, the physical exhaustion that accompanies me.

Similarly, I do not represent the future, the evening I have in front of me as what certainly will come or by making a series of conjectures and fantasies about it; sure, I can do this, but in this case the conjectures or expectations always find their root in feeling, in brief, that something will come, that something else will appear. Finally, from this perspective, one also understands that "the present itself (…), is not posited": it is, rather, what I do in my field of perception, it is my inhering to the surrounding world, my carrying out a task, instead of thinking of it or representing it to myself (Merleau-Ponty 1945: 483).

So, it is the actual existence of my own body that is indispensable to the consciousness of time. In fact, it is only inasmuch as I am a body that inheres in the world, that it lives in situations, that I can conceive of time: in short, we could say, with the philosopher, that 'I exist time'.

But not only that.

As just mentioned, it is thanks to this inherence that the present is not closed in on itself and "the plenitude of being in itself" can be broken: "I am not, for myself, at this very moment, I am also at this morning or at the night which will soon be here, and though my present is, if we wish so to consider it, this instant, it is equally this day, this year or my whole life" (Merleau-Ponty 1945: 489). And again: "I do not form a mental picture of my day, it weighs upon me with all its weight, it is still there, and though I may not recall any detail of it, I have the impending power to do so, I still 'have it in hand'. In the same way, I do not think of the evening to come (…). Ahead of what I see and perceive, there is, it is true, nothing more actually visible, but my world is carried forward by lines of intentionality which trace out in advance at least the style of what is to come" (Merleau-Ponty 1945: 483).

So my present existence is a continuous taking up and anticipation: which is to say that the insertion in the present is access to the whole of time, to the past and the future. This is, in short, a living time together, here and everywhere, now and at all times at the same time. This is why at a certain point, Merleau-Ponty can write: "I am myself time" and add: "We are not saying that time is *for* someone (…). We are saying that time is someone (…). We must understand time as the subject and the subject as time" (Merleau-Ponty 1945: 489 and 490).

Evidently, here, more than a linear unfolding what happens is a dimensional transfer. It is no coincidence that the philosopher claims that there is nothing but a single phenomenon of flowing: "Time is the one single movement appropriate to itself in all its parts" (Merleau-Ponty 1945: 487).

But not only that.

Of utmost importance is to highlight how this flow is the result of a going out of itself, of a centrifugal trend, ecstatic.

This, then, is time's way of being: a single continuous transition from one moment to another, or, rather, a single but uneven wave that is made by differentiation in the transition from the anticipation to the presence and from the new presence to the already happened.

Far from entailing a disintegration it is instead precisely due to this flow that the moments remain, one after the other. We may well say, that they pass and therefore remain. They remain because they pass. So: "in short, since in time being and passing are synonymous, by becoming past, the event does not cease to be" (Merleau-Ponty 1945: 488).

This is a relevant statement, at least to the extent that in this way the structuring of the idea of an objective time is brought back not to an "eternal synthesis, but in the mutual harmonizing and overlapping of past and future through the present, and in the very passing of time" (Merleau-Ponty 1945: 488).

Therefore, not only is the subject temporal and subjectivity and temporality coincide, but it is from this acting intentionality that one can reassert the objectivity of time.

And more.

The same eternity can find sense only starting from the event of this time "as an indivisible and thrust transition" that bears and raises itself.

"The feeling for eternity is a hypocritical one," we read, "for eternity feeds on time".

And so?

"Of what nature, then, is that waking time in which eternity takes root?"

It is the field of presence in which they are located and in which they are engaged, "with its double horizon or primary past and future, and the infinite openness of those fields of presence that have slid by, or are still possible" (Merleau-Ponty 1945: 491 and 492).

Thus, the analysis of the perceptual field has *clarified* the way Merleau-Ponty thinks of time.

It is time to take a last step and try to see how the analysis of time which has emerged so far, *confirms* the co-original emergence of subject and world in the field of perception.

The first movement to be done, then, is to highlight how, from the above description of time, it shows how subjectivity is radically imbued with multiplicity, rooted in the other than oneself.

Well, notes Merleau-Ponty, if the subject is temporality, the self-position, of which he always speaks in reference to subjectivity, will cease to be a contradiction: what time shows in fact, is not a "being which reposes within itself" or a "motionless identity with itself" but rather a subjectivity which is essential "to open itself to an Other and to go forth from itself " (Merleau-Ponty 1945: 495).

This means, in other words, that the analysis of temporality brings us right to the heart of the original relation that every *cogito* is.

Taking up the well-known Kantian expression (Kant 1781 A, 1787 B), Merleau-Ponty notes in fact that this self-position is nothing but a *self-affection*.

Now, to say self-affection means to refer to a subjectivity that feels itself and, in so doing, splits, divides from itself, becomes other than itself: on the one hand, it feels itself as that which feels, suffers—as the subject of the feeling (*celui qui affecté*)—and on the other hand discovers, in his intimate being, the suffering as felt —the object felt (*celui qui est affecté*). In short, writes the philosopher, time as self-affection indicates on the one hand the sentient being, the subject, "as a thrust and a passing towards a future", on the other hand the felt, the object, "as an unfolded series of presents" (Merleau-Ponty 1945: 494).

Husserl thus had good reason when in his *Lectures on the Inner Consciousness of Time* (Husserl 1928) he emphasized how the temporal flux of subjectivity was a flow that is constituted as a *phenomenon in itself*, or a *self-manifestation*.

Radicalizing the Husserlian position, Merleau-Ponty however takes a step further: the intentionality that emerges from his analysis of temporality is not only the place where there is shown the structural relationship that the *cogito* is.

As radically acting intentionality, it shows the original relationship as the site of the occurrence of the self as another, and, correspondingly, the other as oneself.

For this reason, there is no doubt that the analysis of temporality reiterates the co-originality of the I and the world, their mutual intertwining, the birth together, in the field of perception, of the making dull and unreflective of one's own body and of the sense of the world: "It is through temporality that there can be, without contradiction, ipseity, significance and reason" (Merleau-Ponty 1945: 495).

Only thanks to the self-affective dimension of time can there unravel that double movement, the continuous passage that establishes the permanence of the thing on the permanence of my body and the permanence of the *Leib* in the unity of the thing.

Equally clear, however, is that this analysis says something more.

The radical factuality of our spontaneity or, in other words, the passive nature of our every synthesis, in fact, shows well "that we make our way into multiplicity, but that we do not synthesize it." Which means that, as temporal as I am, it is just as obvious is that "I am not the creator of time than any more than of my heart-beats," the same way that "I am not the initiator of the process of temporalization." It is true that "I did not choose to come into the world, yet once I am born, time flows through me, whatever I do."

Yet this flow of time in me is never something that I suffer simply: it is always possible, in fact, "for I can find a remedy against it in itself," make a gesture of freedom (Merleau-Ponty 1945: 496).

Certainly, it will always be a matter of doing in a situation, in things, always radically embodied in the multiplicity of the perceptual field.

From this point of view there is no doubt that for Merleau-Ponty, the subject may be said to be free only if it can be extraneous to itself: that is, to grasp and incorporate the *logos* of the world whose radical otherness we have discovered traverses ourselves. In other words: no donation of sense that the *cogito* shall wish to do, can be exempted from being together a reception of that kind of reversed intentionality that comes from things, events, situations, history, the others.

Thus, far from being the exercise of an absolute entity, freedom is the continuous work of a conditioned corporeity. It is a being in time and not in spite of or beyond it. A birth *into* the world that is always first to be born *from a* world.

References

Brena, G. L. (1969). *La struttura della percezione. Studio su Merleau-Ponty*. Milano: Vita e pensiero.

Heidegger, M. (1927). *Sein und Zeit*. Halle: Niemeyer. English edition: Heidegger, M. (1962). *Being and time* (J. Macquarrie & E. Robinson, Trans.). London: SCM Press.

Husserl, E. (1913). *Ideen zu einer reinen Phänomenologie und phänomenologischen Philosophie. Erstes Buch: Allgemeine Einführung in die reine Phänomenologie*. Halle: Niemeyer Verlag. English edition: Husserl, E. (1982). *Ideas pertaining to a pure phenomenology and to a phenomenological philosophy—first book: General introduction to a pure phenomenology* (F. Kersten, Trans.). The Hague: Nijhoff.

Husserl, E. (1928). *Vorlesungen zur Phänomenologie des inneren Zeitbewusstseins*. Halle: Nyemer. English edition: Husserl, E. (1990). *On the phenomenology of the consciousness of internal time (1893–1917)* (J. B. Rough, Trans.). Dordrecht: Kluwer.

Kant, I. (A 1781, B 1787). *Critik der reinen Vernunft. Riga: Hartknoch*. English edition: Kant, I. (1996). *Critique of pure reason: Unified edition (with all variants from the 1781 and 1787 editions)* (W. S. Pluhar (ed)., J. W. Ellington Trans.). Indianapolis (In.): Hackett Publishing Company.

Koffka, K. (1935). *Principles of Gestalt psychology*. San Diego: Harcourt.

Kölher, W. (1929). *Gestalt psychology*. New York: Liveright.

Merleau-Ponty, M. (1942). *La Structure du comportement*. Paris: Presses Universitaires de France. English edition: Merlaeu-Ponty, M. (1963). *The structure of behavior* (A. Fisher, Trans.). Boston: Beacon Press.

Merleau-Ponty, M. (1945). *Phénoménologie de la perception.* Paris: Gallimard. English edition: Merleau-Ponty, M. (1962). *Phenomenology of perception.* (C. Smith, Trans.). London-New York, Routledge & Kegan Paul.

Schilder, P. F. (1935). *Image and Appearence of Human Body.* London: Kegan Paul, Trench, Trübner & Co.

Spinicci, P. (2000). *Sensazione, percezione, concetto.* Bologna: il Mulino.

Time Out of Joint: Gilles Deleuze and Felix Guattari on Time and Capitalism

Alessandro Arienzo

Abstract Karl Marx theory of value/labour is primarily based on time. In his theory of value/labour, Marx displays how the economic mechanic of Capital reduces Labour to power and time. Power is the ability to produce, and represent a complex mixture of individual workforce and social cooperation. Time is the general measure of productivity and the partition of labour time gives the units of measure of the value produced. Capitalism is driven by one single linear and universal temporality, signed by the time of production, and by the amount of time/value subtracted to the worker. Within an unorthodox Marxist tradition, Gilles Deleuze and Felix Guattari openly criticised this approach to the production of value. By taking their distance from an image of history as a timeline, Deleuze and Guattari sketched history as a geography, and the capitalist society as an archipelago of temporalities. In this chapter, we will discuss their critique of an idea of linear time. Time is a nexus of lines, flows, segmentations and plateau: it is not merely a subjective experience, nor an objective/quantitative measurement of movement. It rather express a cartography of forms of life, of regimes and assemblages.

Karl Marx theory of value/labour is primarily based on time. Workers sell their human labour power to the capitalist who pays for the worker's ability only what they need to reproduce their labour power. The amount of time/labour, which is kept by the capitalist, represents the quantity of workforce extracted to produce plus value. "Absolute plus value" is the result of the increase of value through the extension of the time of labour. "Relative plus value" is produced while keeping fixed the time/labour with an increase in the productivity of the whole process. In this second case, it is the "technical composition" of living labour, namely of variable and fixed capital, to increase the productivity and the capacity to extract a higher quantity of value. In his theory of value/labour, Marx displays how the economic mechanic of Capital reduces labour to power and time. Power is the ability to produce, and represent a complex mixture of individual workforce and social cooperation. Time is the general measure of produc-

A. Arienzo (✉)
Department of Humanities, Philosophy Section, University of Naples Federico II, via Porta di Massa 1, 80133 Naples, Italy
e-mail: alessandro.arienzo@unina.it

F. Santoianni (ed.), *The Concept of Time in Early Twentieth-Century Philosophy*, Studies in Applied Philosophy, Epistemology and Rational Ethics 24,
DOI 10.1007/978-3-319-24895-0_10

tivity and the partition of labour time gives the units of measure of the value produced. Capitalism is driven by one single linear and universal temporality, signed by the time of production and by the amount of time/value subtracted to the worker. Within an unorthodox Marxist tradition, Gilles Deleuze and Felix Guattari openly criticized this approach to the production of value. In *A Thousands Plateaus*, the two philosophers propose a different interpretation of the relation of value/labour characterizing capitalism (Choat 2010). The production of value is not primarily based on time as a quantitative measure, but it is the result of a qualitative regulation of labour force which ultimately results in the production of subjectivities. Labour force is therefore interpreted as a "machinic" structure, and as an assemblage of "different processes of productions", in which a multiplicity of temporalities converge. By taking their distance from an image of history as a timeline, Deleuze and Guattari sketched history as a geography, and the capitalist society as an archipelago of temporalities. Starting from the distinction between *Chronos* on *Aîon* discussed by Deleuze in his earlier works, up to Deleuze and Guattari's later works, this chapter will discuss their critique of an idea of linear time. In their approach, time is a nexus of lines, flows, segmentations and plateau. In this sense, it should not merely a subjective experience, nor an objective/quantitative measurement of movement; it rather expresses a nexus of temporalities that picture a cartography of forms of life, of regimes and assemblages. Capitalism is not a mere process of extraction of value, but is a machinery that produces subjectivities. The countering of Capitalism is therefore the continuous and active capacity to produce forms of life, which do not place themselves in a historical continuity or into a timeline of value production.

Representing time has always been a challenging effort. The most typical depiction of time is that of a continuous line moving from the past towards the future. Time is a measure of motion, and it bears with it the ideas of direction and linearity. Not surprisingly, from antiquity to modern age timelines have always been used to represent and picture historical processes by giving them a rationale, whether it was theological, moral or merely a chronological order. Within this linear representation, time express a flow of events, or a set of choices in a more complex arboreal/genealogical picture.

In their *Cartographies of Time. A History of Timeline*, Daniel Rosenberg and Anthony Grafton investigate the history of the graphic representation of time through *time maps*. Their work is a history of lines and of the attempt to dominate complexity through linearity. From the classic Eusebian model, chronologies and genealogies aimed at assembling valuable information and tied it to memorable graphics. In this sense, timelines express a principle of authority and authoriality that while describing events, prescribe their relations and our relation with them. Seen from a different angle, their volume is also the narration of the effort to escape the linear representation of history, and to elaborate alternative modes of representing the series of events. The authors, in fact, admit that: «our idea of time is so wrapped up with the metaphor of the line that taking them apart seems virtually impossible» (Rosenberg and Grafton 2010: 13). Grafton and Rosenberg thus remind us that in their *Matter and Memory* (1896), Henry Bergson has pictured the metaphor of timeline as a deceiving idol: «an idol of language, a fiction [...] In reality there is no one rhythm of duration; it is possible to imagine many different rhythms which, slower or faster, measure the degree of tension or relaxation of different kinds of consciousness, and thereby fix their respective places in the scale of being» (Bergson 1896: 207). In other words, «space alone is homogeneous; duration and succession belong not to the external world, but to the conscious mind» (Bergson 1896: 120, 1889).

Henri Bergson was an influential philosopher for Gilles Deleuze and Felix Guattari. In their philosophical and political investigation, they question the dominium of linearity and attempt to sketch a plurality of "cartographies" based on a complex topology made of temporal and spatial lines, curves, plateaux as well as processes of *territorialization* and *deterritorialization*. All of them are acting at different molar and molecular levels. Against the sovereign principle of linearity ruling time and history, Deleuze and Guattari propose a different world in which reversibility, discontinuity and indeterminacy are intertwined with moving lines, continuous flows and *machines*.

The latter term is a very complex one. A machine is "a system of interruptions or breaks" (*Anti-oedipus*: 36) or, in other words, a determination or a set of stable relations between flows. Deleuze derives the concept of flow from the writings of the philosopher Henri Bergson and by his theory of multiplicity, and both Deleuze and Guattari widely use it in their works having also in mind the problem of the economy of flows (Deleuze 1971). In his *Cours Vincennes* held in 1971 and dedicated to *A Thousand Plateaus* and to the *Anti-Oedipus* Deleuze points out how the concept of flux cannot be understood by posing the question of the nature of the things flowing in it. Rather, the flux is the correlate of five other notions: pole, code or accounting system, stage of transformation, sector and stock. In other words, the flux can be better understood by pointing out the break flow. The flux is the pure movement of thing that is at the same time necessary and made possible by the existence of differences (the poles), and accounting system (a coherent system of passages between the poles), a process of transformation between the two poles that are also poles of concentration or scarcity (sector and stock). Deleuze, thus, clarifies that

> For this notion of the break-flow has to be understood simultaneously in two ways: it is to be understood as the very correlation of flux and code, and if, returning once more to capitalism, we are aware that flows are "accounted for", it is in favour of a movement of decoding such that the accounting system has simply taken the place of codes; it is at this point that we come to realize that it's no longer sufficient to speak of an accounting system, but rather of a financing system or structure (Deleuze 1971).

In this sense, a society is always constituted by flows, and a person is always a cutting-off (*coupure*) of a flow. Moreover, at the same time, a person "is always a point of departure for the production of a flow, a point of destination for the reception of a flow, a flow of any kind; or, better yet, an interception of many flows" (Deleuze 1971, 2003). Thus, a machine is "a form" that, for this simple reason, constitutes a break, a *couture*, in a plain of consistency. The philosophy of Deleuze and Guattari is clearly a philosophy of immanence and the "plain of consistency": "Far from reducing the multiplicities number of dimensions to two, the *plane of consistency* cuts across them all, intersects them in order to bring into coexistence any number of multiplicities, with any number of dimensions. The plane of consistency is the intersection of all concrete forms" (Deleuze and Guattari 1980: 251; Ciccarelli 2008). The concept of "machine" expresses the networking of a plurality of lines, flows, curves (of spaces as well as temporalities) in which the productive capacity of desire realize itself. The holding together of diverse element is a consistency and it represents a *style* of existence.

This immanent philosophy and the concept of flow are already present in Gilles Deleuze earlier works, in particular in his *Difference and Repetition* (1968), *Logic of Sense* (1969), and *Bergsonism* (1966). In the former volume, Deleuze confronts with Kantian concept of synthesis and discusses *three syntheses of time*. While in Kant syntheses are activities undertaken by the mind or the subject, in Deleuze they are passive processes that are constitutive of both minds and subjects: "Every determinate thing is a combination of singularities, forming a multiplicity that is changing in multiple ways according to the syntheses of time" (Williams 2011: 187n). The first, passive, synthesis has an organic nature and is a *contracted habit* being the living present of the body and a *mens momentanea* in which the past is our genetic heredity and the future is a mere necessity. The second synthesis is the "memory", and it resembles the Bergsonian concept of a pure past. The third synthesis concerns the conditions for the production of the new and shows that: "time out of joint means demented time or time outside the curve which gave it a god, liberated from its overly simple circular figure, freed from the events which made up its content, its relation to movement overturned; in short, time presenting itself as an empty and pure form. Time itself unfolds (that is, apparently ceases to be a circle) instead of things unfolding within it (following the overly simple circular figure). It ceases to be cardinal and becomes ordinal, a pure order of time" (Deleuze 1968: 88).

In his *Logic of Sense* there is a further distinction between *Chronos* and *Aîon* that he derived from the stoics. In broad terms, Chronos represents the chronological time. In Chronos only the present "exists", and past and future are its extensions under the figure of motion. Chronos is "form", it is the development of a form. Deleuze alse describe Chronos as a pulsed time, not necessarily regular or periodic, that "punctuates" the formation of a subject. Aîon is a completely different time, which the Stoics described as proper of the incorporeal. Aîon only subsist only in the past and in the future as in it the present has no other existence than being an instant that lean towards the past or the present. In Aîon the instant is never "present" to itself. Nonetheless, Deleuze places in its evanescence and openness to future the grounds of language and mutation. Within this topology characterized by the intertwining of Chronos and Aîon, time and movement are expressed in a very peculiar way: as "becoming different", becoming woman, becoming revolutionary, becoming philosophy.

Time is not a measure of movement but is in itself "motion": the motion of desire (Deleuze and Guattari 1983, 1986, 1989). That of the desire is a fundamental concept in Deleuze and Guattari. Against psychoanalysis, and against the structuring of the unconscious as the theatre of the Oedipal tragedy, or as the locus of an absence (Deleuze and Guattari 1972, 1980), Deleuze and Guattari develop the idea that the unconscious is a desirous machine and that desire is production. Psychoanalysis defined the unconscious moving from the ideas of "absence" and "want". Civilization is based on this domination of libido and instincts, and on the capacity to renounce and "sublimate" the instincts. The correlate of renunciation is "the Law", barring the subject from reaching the object of its desire. In fact, desire is always the desire of "a Thing" that must be barred to promote the transference of the instinctual aims.

Against this "reactive" view of desire—desire of something absent, wanting or lost—Deleuze and Guattari declared that desire does not depend on missing "a Thing" and it is not a prelude to any law. Desire is production. Any social arrangement has its matrix in desire and in the productive capacity of the unconscious. In fact, the unconscious is in itself a social and collective arrangement, and is continuously traversed by desires that are never exclusively individual. Capitalism is not merely a social system, but is a specific libidinal economy, developing through the investments in cash flows, means of production, markets and commodities. These are all assemblages within an economy of desire.

The Anti-oedipus and *A Thousand Plateaus* are the two moments of an intellectual effort that was entitled by Deleuze and Guattari *Capitalism and Schizophrenia*, for the very reason that capitalist economy function in a continuous circle from *de-territorialisation* (dis-individuation) to *territorialisation* (individuation) and back again. In this sense, interpreting capitalism takes the same effort of interpreting psychosis, more specifically schizophrenia. In fact, schizophrenia has the capacity to realize—and potentially does realize—all the different psychotic forms: in a very peculiar way, it is a unity of differences. Thus, capitalism resembles the schizophrenics in its capacity to permanently *de-codify* and *de-territorialise* itself to the limit, and transform itself in contrasting options, forms and desires.

The ideas of territorialisation and de-territorialisation bear with them the image of a cartography, and Deleuze and Guattari constantly use images such as lands, borders, lines, planes, curves. These images are both geographical and conceptual, as territoriality is a movement of subjectivisation and individuation while de-territorialisation is the opposite strife for de-individuation and for the production of new arrangement and different "machines".

This interpretation of capitalism as a libidinal economy openly contrasts with those of Freudianism and with Marxism, and clashes frontally with the attempt to merge the two. While Freudianism was criticized for the imposition of an oedipal theatre that was, in fact, a representation of the bourgeoisie family, a major critique was also moved against the Marxist theory of value/labour. According to Marx, workers sell their human labour power to the capitalist who pays for the worker's ability only what they need to reproduce their labour power. The amount of time/labour, which is kept by the capitalist, represents the quantity of the workforce extracted to produce plus value. "Absolute plus value" is the result of the increase of value through the extension of the time of labour. "Relative plus value" is produced while keeping fixed the time/labour with an increase in the productivity of the whole process. In this second case, it is the "technical composition" of living labour, namely of variable and fixed capital, that increases the productivity and the capacity to extract a higher quantity of value. In his theory of value/labour, Marx displays how the economic mechanic of Capital reduces labour to power and time.

Power is the ability to produce, and represents a complex mixture of individual workforce and social cooperation. Time is the general measure of productivity and the partition of labour time gives the units of measure of the value produced. Capitalism is driven by one single linear and universal temporality, punctuated by the time of production, and by the amount of time/value subtracted to the worker. The discipline of Capital is essentially the discipline of time as it has been magisterially described in the works by the historian, Thompson (1967) and recently discussed by Moishe Postone in his *Time, Labor and Social Domination* (1993).

In *A Thousands Plateaus*, the two philosophers propose a different interpretation of the relation of value/labour characterizing capitalism. The production of value is not primarily based on time as a quantitative measure, but is the result of a qualitative regulation of labour force, which ultimately results in the production of subjectivities. Labour force is therefore interpreted as a "machinic" structure, and as an assemblage of "different processes of productions", in which a multiplicity of temporalities converge. Capitalism is not a mere process of extraction of value, but is a machinery that produces subjectivities and individuation.

In this sense, capitalism is the radical decoding and deterritorialization of the flows that previous social machines had zealously coded, namely the feudal society. Indeed, capitalism is also a coding and territorializing social machine that connects deterritorialized flows of labour and capital and extracts a surplus from that connection. Thus capitalism sets loose an enormous productive charge the surpluses of which is captured and controlled by the institutions of private property that associate this production to individuals. Those individuals are not "natural" as they have a social nature which is, in fact, a social division between capitalist or labourer and, in a second instance, private as they belong to specific family. In Deleuze and Guattari's terms, capitalism's decoded flows are reterritorialized on "individuals" with their correlate of individual rights—the most important is that of private property—and a psychological configuration as they are family members as figures in the Oedipal triangle (Toscano 2006).

The countering of Capitalism is therefore the continuous and active capacity to produce forms of life, which do not place themselves in a historical continuity or into a timeline of value production. Schizoanalysis deals with the continuous search for the condition of a revolutionary political struggle (Guattari 1989a, b). However, revolution is *nomadism*, i.e. the transversal freeing of desire in its productive nature, and in its capacity for becoming something else. By taking their distance from an image of history as a timeline, Deleuze and Guattari sketch history as a geography, and the capitalist society as an archipelago of temporalities. In their approach, time is a nexus of lines, flows, segmentations and *plateaus*. In this sense, it should not be merely a subjective experience, nor an objective/quantitative measurement of movement; it rather expresses a cartography of forms of life, of regimes and assemblages always in becoming. Time is not a measure but a quality and, in this sense, it has more to do with geography than with history. Reality is [in] becoming and becoming is geographical for things and people are made of different and uncertain "lines, directions, entrances and exits".

Time is primarily collective, multiple and differentiated, and can be traced in the qualitative punctuation of forms in their becoming different (woman, philosophy, revolutionary). It is a temporal stream of consciousness whose nature is cooperative and trans-individual (Williams 2011). Against the conservatism of psychoanalysis, schizoanalysis is part of a revolutionary struggle aiming at the liberation of fluxes from the Superego (Guattari 1989, 2004, 2008, 2010). Against the orthodoxy of a Marxism based on a systematic timeline of dominations and revolutions preparing the communism to come, Deleuze and Guattari affirm that the revolution is not an act or a historical event; it is rather a "becoming different" through a desiring economy alternative to that of capitalism.

References

Bergson, H. (1889). *Essai sur les données immédiates de la conscience*. Paris: Alcan. English edition: Bergson, H. (1910). *Time and free will: An essay on the immediate data of consciousness* (F. L. Pogson, Trans.). London: George Allen & Unwin.

Bergson, H. (1896). *Matière et mémoire: essai sur la relation du corps à l'esprit*. Paris: Alcan. English edition: Bergson, H. (1988). *Matter and memory* (N. M. Paul & W. S. Palmer, Trans.). New York: Zen Books.

Choat, S. (2010). *Marx through post-structuralism. Lyotard, Derrida, Foucault, Deleuze*. London: Continuum Books.

Ciccarelli, R. (2008). *Immanenza. Filosofia, diritto e politica della vita dal XIX al XX secolo*. Bologna: Il Mulino.

Deleuze, G. (1968). *Différence and repetition*, Paris: Presses Universitaires de France. English edition: Deleuze, G. (1994). *Difference and repetition* (P. Patton, Trans.). New York: Columbia University Press.

Deleuze, G. (1969). *Logique du sense*. Paris: Les Editions de Minuit. English edition: Deleuze, G. (1990). *Logic of sense*. (Boundas, C. V. (Ed.), Lester, M. Trans.). New York: Columbia University Press.

Deleuze, G. (1971). Anti Oedipe et Mille Plateaux. Nature des flux. *Cours Vincennes* 14 Dec http://www.webdeleuze.com/php/sommaire.html.

Deleuze, G. (1983). *L'Image-mouvement. Cinéma 1*. Paris: Les Éditions de Minuit. English edition: Deleuze, G. (1986). *Cinema 1: The movement image* (Habberjam, B. Trans.). Minneapolis: University of Minnesota Press.

Deleuze, G. (1985). *L'Image-temps. Cinéma 2*. Paris: Les Éditions de Minuit. English editon: Deleuze, G. (1989). *Cinema 2: The time image* (B. Habberjam, Trans.). Minneapolis: University of Minnesota Press.

Deleuze, G. (2003). *Deux Régimes de Fous, Texte et entretiens 1975–1995*. Paris: Les Éditions de Minuit.

Deleuze, G., Guattari, F. (1972). *L'Anti-oedipe*. Paris: Les Editions de Minuit. English edition: Deleuze, G., Guattari, F. (2004). *The anti-oedipus* (R. Hurley, M. Seem & H. R. Lane, Trans.). London, New York: Continuum. Deleuze, G., Guattari, F. (1983). *On the line* (J. Johnston, Trans.). Los Angeles: Semiotext(e).

Deleuze, G., Guattari, F. (1980). *Mill-plateaux*. Paris: Les Editions de Minuit. English edition: Deleuze, G., Guattari, F. (2004). *A thousand plateaus. Capitalism and Schizophrenia*. (B. Massoumi, Trans.) London, New York: Continuum.

Deleuze, G. (1966). *Le Bergsonisme*. Paris: Puf. English edition: Bergsonism. (1991). (H. Tomlinson, & B. Habberjam, Tans.). New York: Zone Books.

Guattari, F. (1989a). *Cartographies Schizoanalytiques*. Paris: Galilée.

Guattari, F. (1989b). *Le trois écologies*. Paris: Galilée. English edition: Guattari, F. (2000). *The three ecologies* (I. Pindar & P. Sutton, Trans.). London, New Brunswick, New York: The Athone Press.
Guattari, F. (1992). *Chaosmose*. Paris: Galilée. English edition: Guattari, F. (1995). *Chaosmosis. An ethico-aesthetic paradigm* (P. Bains, & J. Befanis, Trans.). Bloomington, Indianapolis: Indiana University Press.
Guattari, F. (2004). *The anti-oedipus papers*. Semiotext(e): Los Angeles.
Guattari, F. (2008). *Chaosophy, new edition: Texts and interviews 1972–1977*. Semiotext(e): Los Angeles.
Guattari, F. (2010). *The machinic uncoscious. Essays in schizo-analysis*. Semiotext(e): Los Angeles.
Postone, M. (1993). *Time, labor and social domination: A reinterpretation of marx's critical theory*. New York, Cambridge: Cambridge University Press.
Rosenberg, D., & Grafton, A. (2010). *Cartographies of time. A history of timeline*. Princeton: Princeton Architectural Press.
Thompson, E. P. (1967). Time, work-discipline and industrial capitalism. *Past and Present, 38*(1), 56–97.
Toscano, A. (2006). *The theatre of production. philosophy and individuation between Kant and Deleuze*. Basingstoke: Palgrave Macmillan.
Williams, J. (2011). *Gilles Deleuze's philosophy of time: A critical introduction and guide*. Edimburgh: Edinburgh University Press.

On Derrida's Critique of the Metaphysics of Presence. Implications for Scientific Inquiry

Marco Stimolo

Abstract The chapter delivers a biref description of Derrida's critique of the metaphisics of presence and shows its relevant implications for scientific inquiry. It is shown that the deconstruction of the concept of Being as a trascendental signifier is consistent with the scientific use of operational definitions that simplify the complexity of real world phenomena through the use of unrealistic assumptions. The argument reaches two main conclusions. First, the deconstruction of the trascendental signifier entails that unrealistic assuptions do not have a metaphysical status and they can change in accordance with the specific purposes of scientific inquiry. Second, the decostrustionistic framework implies that scientific analysis can asymptotically approximate the target phenomenon without reaching a full correspondence with it. These results are illustrated through the example of economic analysis of individual behaviour.

This chapter delivers a brief review of Derrida's deconstruction of the western metaphysics of presence. On these grounds, the study puts emphasis on the coherence of Derrida's critique of the metaphysical tradition with the real practice of scientific inquiry, with particular concern to the economic analysis of individual behavior. The chapter starts by illustrating the basic features of the metaphysics of presence through two paradigmatic examples: Aristotle and Husserl. These philosophical inquiries share the common feature of assuming the existence of a transcendental signifier that justifies a conception of truth as *adaequatio mentis ad rem*. The chapter illustrates the main implications of the deconstruction of the metaphysics of presence for scientific inquiry. More precisely, it explains how the supposed existence of a transcendental signifier is inconsistent with the scientific use of operational definitions that change according to specific analytical purposes. Indeed, operational definitions ground in *simplifying* assumptions that are necessary for the analytical tractability of the phenomenon of interest. The use of simplifying assumptions entails the *incompleteness* of the result of scientific inquiry. In this

M. Stimolo (✉)
Department of Humanities, Philosophy Section, University of Naples Federico II, via Porta di Massa 1, 80133 Naples, Italy
e-mail: marcostimolo@virgilio.it

F. Santoianni (ed.), *The Concept of Time in Early Twentieth-Century Philosophy*, Studies in Applied Philosophy, Epistemology and Rational Ethics 24,
DOI 10.1007/978-3-319-24895-0_11

regard, the proposed argument highlights the coherence of this methodology with the deconstruction of the transcendental signifier that extends asymptotically the role of scientific analysis. The chapter is organized as follows. Section *The metaphysics of presence* describes the basic features of the metaphysics of presence. Section *Deconstruction of the metaphysics of presence. Basic implications for scientific inquiry* identifies the implications of Derrida's deconstruction of the transcendental signifier for scientific inquiry. Section *An illustrative example* illustrates these implications through the example of the economic analysis of individual behavior. Final section concludes.

1 The Metaphysics of Presence

«The entire history of the concept of structure…. must be thought of as a series of substitutions of center for center…. The history of metaphysics… is the history of these metaphors and metonymies. Its matrix… is the determination of Being as presence in all senses of this word. It could be shown that all names related to fundamentals, to principles, or to the center have always designated an invariable presence—eidos, arche, energeia, ousia (essence, existence, substance, subject)» (Derrida 1967a, b: 279–280).

This long quotation provides us the fundamental features of Derrida's understanding of the history of the western metaphysics. The grounding idea of this passage is that, beyond the differences of all the metonymic substitutions of center for center that accumulated through the centuries, western metaphysics exhibits the invariant feature of thinking of Being in terms of presence. Indeed, the Being can be conceived in terms of idea, *ousia*, God, and so on, but what holds together all of these different concepts is the *modality* of thinking about it as a presence.

The term "presence" means that both human thought and its language refer to something external, which does not change independently of the variation of its conceptual and linguistic expressions. In Derrida's view, the Being is a "transcendental signifier", which is intended to be the eternal and final reference of the discourse, writing, and inquiry (Derrida 1967a, b, 1973). In other words, the transcendental signifier is the metaphysical foundation of epistemic certainty: what humans get to know through scientific inquiry is ultimately guaranteed by Being thought as a presence.

According to Derrida, such metaphysical stance entails the utopia of eliminating the linguistic mediation from human understanding of reality. On this account, signs and words play the role of mediators between the subject and the objective reality, but they are eventually eliminated when the subject corresponds completely to its object.

These general features of the metaphysics of presence are detectable in qualified forms in two paradigmatic examples: Aristotle and Husserl.

According to Derrida (1976), Aristotle in the *Organon* (Categories) aimed at identifying the place where the thought/language coupling emerges. In this theoretical framework, the Being is intended to be the most general category (i.e., the category of categories) that opens language to its external reference, which *is not linguistic in itself*. Hence, the Being guarantees an ontological foundation to all the categories. Indeed, Aristotle's theory of the categories pursues the objective of describing and determining the different enunciates through which Being can be analyzed. However, any enunciate assigns a predicate to Being that guarantees their ontological foundation. On this account, the linguistic mediation ceases whenever the subjective knowledge corresponds completely to Being.

Modern epistemology and Cartesian philosophy in particular assume a dualism between psychic objects and some material objects existing outside in the real world. These basic features characterize Husserl's phenomenology, which Derrida considers the completion of the Cartesian tradition (Derrida 1973). Husserl's phenomenology aims at identifying the metaphysical foundation (i.e., the transcendental signifier) within the structure of subjectivity (Husserl 1931). In Derrida's view, the ultimate goal of Husserl's phenomenology is the identification of the purely rational foundation of philosophy in the ideas that consciousness generates. Husserl inscribes a dualism between the purely formal (empty) subjective intention (i.e., noesis) and the objective intuition (i.e., noema). The formal subjective intention mediates between the transcendental consciousness and its objects. However, the necessary truth of the objects of consciousness (i.e., noemata) emerges only if they fully correspond to the objects previously given in the empty intention (i.e., noesis). Therefore, all the mediating signs of language fade away whenever the noesis is in the immediate presence of its noematic intuition.

These examples are paradigmatic because they show how the switch from classical ontology (Aristotle) to modern epistemology (Cartesian philosophy and Husserlian phenomenology) did not affect the general feature of the metaphysics of presence. Indeed, in both cases Being is thought as a presence that grounds the certainty of human knowledge and in the end eliminates the very need of thought itself.

2 Deconstruction of the Metaphysics of Presence: Basic Implications for Scientific Inquiry

The basic feature of Derrida's critical project consists in the attempt to undermine the possibility of eliminating the linguistic mediation by deconstructing the transcendental signifier (Derrida 1967a, b, 1973). Given the intrinsic complexity of Derrida's critical project, the proposed argument exclusively focuses on the implications of deconstruction for scientific inquiry.

Consider the main objective of science: explaining a phenomenon by identifying its cause. Real phenomena exhibit a high degree of variability that is not analytically tractable. This means that the noncontrollable variability of a phenomenon is a confounder for causal attributions. To avoid this problem, scientific inquiry grounds in operational definitions of the target phenomenon. This particular kind of definition *holds constant* some aspect of the phenomenon according to the specific purposes of the analysis (Garrison 1999). Through this expedient, scientific inquiry obtains clear results that are useful to interpret real-world phenomena.

In Derrida's view, the grounding mistake of western metaphysics consists of two false beliefs. First, the western metaphysics believed that outside the knowing subject there exists an *unchanging* referent. As far as scientific inquiry is concerned, this false belief requires that the basic assumptions and the results of the analysis have an immediate referent in the real world. Second, according to western metaphysics, the transcendental signifier is an absolute center that scientific inquiry can approximate through a convergent dynamic. On this account, scientific analysis can aim at a perfect correspondence with its object so as to eliminate any analytical mediation.

These implications of the metaphysics of presence prove to be inadequate to the real practice of science. The next section illustrates such an inadequacy through the example of the economic analysis of individual behavior.

3 An Illustrative Example

Derrida's deconstruction of the transcendental signifier has two crucial implications for scientific inquiry. First, the inference from the basic assumption of the analysis to an external referent is fallacious. Second, the inference from the result of a scientific inquiry toward an external referent in the real world is incomplete because the *difference* between the external referent and its analytical expression can never be overcome. This means that scientific inquiry is endless for it can never reach a perfect correspondence with its object.

To illustrate the point, consider the economic definition of a rational individual: an individual is rational to the extent that (s)he maximizes his or her utility function. This definition is operational because it allows one to measure the degree of rationality of an individual. However, individuals' behavior in the real world is highly variable because of an indefinite number of influencing factors. Therefore, economic analysis needs to hold something constant to identify general regularities in behavior. To pursue this aim, individuals are assumed to have a set of well-defined preferences that does not change over time. Such an assumption serves the purpose of explaining the average individual behavior in a population. Although the assumption of unchanging preferences is evidently unrealistic, it simplifies the analysis of individual behavior (Mäki 2011).

Returning to Derrida, the crucial fallacy of philosophy is to confuse the idealized product of inquiry as an antecedent metaphysical state of affairs. In the proposed example, the philosophical fallacy would be to consider the assumption of a well-defined and unchanging set of preferences as a metaphysical truth and not as a methodological necessity for an operational definition.

Such a distinction between the methodological status of the assumption of a constant and the philosophical fallacy of retaining it as a metaphysical truth traces back to Nietzsche (2011). According to the philosopher from Rocken, logical thinking and inference are grounded in the condition of assuming the existence of identical cases. If this assumption is not considered as fictitiously fulfilled, any causal inference would be impossible. However, the need of a false assumption does not allow one to retain it to be a metaphysical truth.

If economic analysis of individual behavior grounds in a simplifying assumption of unchanging preferences, then the result of the inquiry can never be a mirror image of real-world phenomena. Indeed, the assumption at stake clears out several aspects of individual behavior. Therefore, the extension of the results of a scientific inquiry to the real behavior of individuals is *incomplete*. To put it in philosophical terms, the full correspondence between the results of scientific inquiries and real-world phenomena is impossible in principle. This is consistent with Derrida's concept of an ineludible difference between linguistic signs and their referent.

In the attempt to account for the intrinsic complexity of real-world phenomena, scientific inquiries need to change or modify the basic assumptions of their operational definitions consistently with the target phenomenon. It is worth noting that only the absence of a transcendental signifier opens up the possibility of changing the operational definition according to the phenomenon of interest. Indeed, within the framework of the metaphysics of presence, the supposed existence of a transcendental signifier would provide a metaphysical status to the grounding assumptions of operational definition, so as to exclude the possibility of changing them according to an instrumental logic.

To illustrate the point, consider a change of the purpose of the economic analysis from the individual behavior *in general* to the analysis of the *specific* effect of peer pressure on individuals' economic choices. In this case, if the assumption of unchanging preferences were a metaphysical truth, we would not observe any conformist behavior, because individuals' preferences would be completely insulated from peer pressure; as a result, individuals would follow *only their own* tastes. Nonetheless, conformism is a widely observed phenomenon that calls for an explanation. Therefore, given the different purpose of the analysis, economic inquiry needs to relax its basic assumption about preferences, allowing them to *change systematically* according to the specific features of the economic environment (Bowels 1998). Thus, the change in the target and purpose of the analysis justifies a change in the basic assumption. This change permits us to construct economic models where the choice behavior of peers affects individual preferences so as to induce a conformist behavior.

The possibility of relaxing the assumptions of operational definition significantly extends the domain and explanatory depth of the economic analysis. Since it allows us to switch from the interpretation of individual economic behavior *in general* to the *specific case* of conformist behavior. However, given the methodological status of this assumption, the inference from the analytical result to the real-world phenomenon is incomplete. This entails that results of scientific inquiry are never identical to their referent. To put it in a deconstructionist language, scientific inquiries do not converge toward an absolute center, where all linguistic and analytical mediations fade away. It follows that scientific inquiries approximate *asymptotically* their target phenomenon, without reaching a perfect correspondence to a transcendental signifier.

4 Conclusions

This chapter has briefly illustrated how the basic features of the metaphysics of presence entail the philosophical fallacy of retaining the idealized product of inquiry as an antecedent metaphysical state of affairs. Furthermore, the proposed argument has shown that this fallacy is inadequate to the real practice of scientific inquiry that grounds in an extensive use of operational definitions. To illustrate the point, the case of the economic analysis of individual behavior has been considered. The paper focused on the methodological need to change the basic assumptions about preferences according to the specific purposes of the inquiry. This practice entails that the results of scientific investigation can never reach a perfect correspondence with their target phenomenon. By virtue of that fact, the argument emphasized the consistency of this methodology with Derrida's deconstruction of the transcendental signifier that extends asymptotically the role of scientific analysis.

References

Bowles, S. (1998). Endogenous preferences: The cultural consequences of markets and other economic institutions. *Journal of Economic Literature, 36*, 75–111.

Derrida, J. (1967a). *De la grammatologie*. Paris: Les Éditions di Minuit. English edition: Derrida, J. 1998. Of Grammatology (trans: Spivak, G. C.). Baltimore: JHU Press.

Derrida, J. (1967b). *L'Écriture et la différence*. Paris: Seuil. English edition: Derrida, J. (1978). *Writing and difference* (A. Bass, Trans.). Chicago: University of Chicago Press.

Derrida, J. (1973). *Speech and phenomena, and other essays on Husserl's Theory of Signs* (D. B. Allison, Trans.). Evanston: Northwestern University Press.

Derrida, J. (1976). The supplement of Copula: Philosophy before Linguistics. *The Georgia Review, 30*, 527–564.

Garrison, J. (1999). John Dewey, Jacques Derrida, and the metaphysics of presence. *Transactions of the Charles S. Peirce Society, 35*(2), 346–372.

Husserl, E. (1931). *Méditations cartésiennes: Introduction à la phénoménologie* (G. Peiffer, & E. Levinas, Trans.). Paris: Vrin. English edition: Husserl, E. (1960). *Cartesian meditations* (D. Cairns, Trans.). The Hague: Nijhof.

Mäki, U. (2011). The truth of false idealizations in modeling. In P. Humphrey & C. Imbert (Eds.), *Representation, models and simulations* (pp. 216–233). London: Routledge.

Nietzsche, F. (2011). *The will to power* (Trans. R. J. Hollingdale, & W. Kaufmann, Ed. W. Kaufamann). New York: Random House LLC.

Commentary: The Phenomenology and Perception of Time

Akiko M. Frischhut

Abstract How are we aware of time? How do we perceive change and duration? What is it like to experience temporality as opposed to spatiality? Does the way we experience time tell us anything about the nature of time? This chapter focusses on some of the most pertinent questions in the philosophy of time—on the relation between subjective and objective time, on the metaphysical and psychological priority of the present, on the phenomenal difference between our experiences of space and our experiences of time, and on how to reconcile relativistic time concepts with our common-sense ideas. It gives us an interesting in sight into leading debates on the topic, largely within the phenomenological and hermeneutic philosophical traditions. This article intends to complement the chapter by providing a brief overview over views and debates about time consciousness in contemporary analytic philosophy.

How are we aware of time? How do we perceive change and duration? What is it like to experience temporality as opposed to spatiality? Does the way we experience time tell us anything about the nature of time?

This chapter focusses on some of the most pertinent questions in the philosophy of time—on the relation between subjective and objective time, on the metaphysical and psychological priority of the present, on the phenomenal difference between our experiences of space and our experiences of time, and on how to reconcile relativistic time concepts with our common-sense ideas. In the contemporary philosophical landscape, as diverse as it is in its different traditions and approaches, questions about time and time consciousness take a central position. The articles you will read in this chapter might be characterized as predominantly situated within the phenomenological or hermeneutical "continental" traditions of philosophy. This article intends to complement the chapter with a brief overview over the relevant debates in contemporary analytic literature on time consciousness and temporal perception.

A.M. Frischhut (✉)
Department of Philosophy, University of Geneva, 2 Rue de Candolle,
1211 Geneva 4, Switzerland
e-mail: akiko.frischhut@gmail.com

F. Santoianni (ed.), *The Concept of Time in Early Twentieth-Century Philosophy*,
Studies in Applied Philosophy, Epistemology and Rational Ethics 24,
DOI 10.1007/978-3-319-24895-0_12

1 The Problem of Temporal Experience

We appear to be no less acquainted with the temporal structure of the world than we are with its spatial structure. But how do we experience time? One plausible answer is that we experience time by virtue of experiencing change, a theory which earns its credibility through the fact that we measure durations of time by observing change. How could the experience of change relate to the experience of time? Here is one answer. A change is always a change in objects, as well as a change of property instantiations at times. Call these property instantiations (as in 'being green at tn) 'atomic events'. When we see a chameleon changing from red to green say, we experience that change as a succession of two events, an event 'chameleon is green at t1', followed by another event 'chameleon is red at t2', and we are thereby aware of the interval (t1–t2) over which the change occurs. Thus, in experiencing change, a subject experiences a succession of atomic events. A succession takes time. It comes at no surprise then that one can infer duration from one's experiences of change, for there could not be change without duration. Experiencing change entails experiencing the duration over which the change occurs, just as experiencing some material object entails experiencing the region of space that is occupied by that very object. And duration just is temporal extension: thus, we experience time by virtue of experiencing change. In experiencing change, we appear to be no less acquainted with the temporal structure of the world than we are with its spatial structure. As John Foster puts it,

> […] duration and change through time seem to be presented to us with the same phenomenal immediacy as homogeneity and variation of colour through space (Foster 1982).

But while our awareness of time and change seems obvious, the analysis of temporal experience encounters an intricate problem: temporal phenomena such as change and duration take time, while our perceptual experiences seem confined to the momentary present. We cannot perceive what has happened or will happen—all we can be aware of in perceptual experience is what is happening now.[1] Let me expand on this briefly. Nobody doubts that we are in *some* way aware of change. One might, for example think that to experience change, for example the change of a chameleon from green to yellow, is to see that the chameleon is yellow, *remember*

[1]Strictly speaking it is not true that we only perceive what is momentarily present. In fact, all we ever perceive is the past, due to the time lag in perceptual experience. The point is, however, that we only ever perceive *a* present (rather than *the* present)—we perceive what *was* present at the time when the light was emitted from the object we are currently perceiving, but we do not perceive *more* than what was *then* present. Our awareness, it seems, is restricted to what happens *at* a time, and cannot 'take in' what occurs *over* time. That is to say, it is natural to assume that when a perceptual stimulus takes *n* seconds to reach us at a time *t*, then we are perceptually aware of what happens at time *t-n*, but not of anything that happened before or after *t-n*. The initial problem thus persists: if we are only ever perceptually aware of what was momentarily present, and change and duration are temporally extended phenomena, then we cannot be perceptually aware of change (or any other temporally extended phenomena).

that it was green and thereby be aware of the chameleon's change—but in that case one would not perceptually experience the change of the chameleon. One would *infer* that the chameleon has changed from memory and experience. Or one might think that to experience change is to be in a composite mental state constituted by a perceptual experience with the content 'the chameleon is yellow' and a memory state with the content 'the chameleon was green', where the combination of both results in an experience of change. Alternatively one might think that one could judge that the chameleon was green, while perceiving that it is yellow or imagine that it was green, while perceiving that it is yellow, where in both cases the combination of the imaginative state, or the judgement with the perceptual state would result in an experience of change. These kinds of more "loosely understood" experiences are not at issue here. The dispute in the debate about temporal perception is whether one can be aware of change *solely by virtue of one's current perceptual experience.* Thus, to be precise, what we are aware of *solely by virtue of our current perceptual experience* seems to be constrained to the momentary present. Here now emerges the philosophical problem: If we can only be perceptually aware of what is (more or less here and) now, at the present moment, then we cannot be perceptually aware of change or duration (or any other temporally extended structure). In other words, if our perceptual awareness is confined to the (or a) present, it must lack any temporal depth. If it lacks temporal depth, then it is impossible for us to perceptually experience change and thus to be perceptually aware of time.

So on the one hand, it seems to us that we perceive change and duration just as we perceive colours and shapes, but on the other hand, it seems to us that we cannot be perceptually aware of more than what is momentarily present. In brief, we are confronted with what some people have called the *paradox of temporal awareness*:

> It seems that, in order to experience any temporally extended phenomena our experiential awareness must extend over time, but it seems that it can't (Dainton 2010).

Here is how Dan Zahavi puts the problem:

> Pre-theoretically we all assume that we have direct experiences of change and persistence. We can hear an enduring tone or a melody, just as we can see a stationary pyramid or the flight of a bird. However, if I at any given moment were only aware of what was perceptually present then and there, how could I then ever perceive—in contradiction to remember, imagine or judge about—temporally extended objects? (Zahavi 2007).

It is an almost universally accepted truth within the literature on temporal perception that the mere fact that a subject S has had an auditory experience of tone *C* at *t*1, say, before having an experience of tone *D* at *t*2, does *not suffice* for S to have an experience of the succession *C-D*. As Husserl puts it,

> [t]he duration of sensation and the sensation of duration are different. And it is the same with [succession]. The succession of sensations and the sensation of succession are not the same (Husserl 1905, 1964: 31).

William James famously expressed the same thought:

> A succession of feelings, in and of itself, is not a feeling of succession: we must think of A and B as one after each other, but we must think them simultaneously (James 1890: 629).

What Husserl and James have in mind is that in order to hear a tune as a tune at any time t, one needs to have an experience of a succession of tones *at that time t*. More generally put, the claim is that to be at any time perceptually aware of a temporal structure such as change, succession or duration, the whole structure must be represented at that very moment in time (Phillips 2010).

In order to have a perceptual experience of a tune, or of a change or a succession in general, all tones, or all parts of the change or of the succession, have to be experienced *together*. The claim that in order to perceptually experience change *as* change, we have to be, *in some way or other*, aware of more than what presently occurs is acknowledged by all sides of the debate. The big question is *how* we are aware of things that do not presently occur. The first major divide in the discussion is between those philosophers that reject the idea that we can perceptually experience change and those which reject the idea that we can only perceptually experience what is momentarily present. I call the former anti-realists about temporal perception (short: anti-realists), and the latter realists about temporal perception (short: realists).[2]

2 The Memory Theory of Temporal Perception

Anti-realists about temporal perception think that if changes were experienced, they would have to be experienced as taking time, and that our perceptual awareness is bound to the present moment. As a consequence, they deny that we can perceptually experience change.

Somebody who famously held this position was Thomas Reid, who argued that we cannot, 'strictly and philosophically speaking', experience change or any kind of succession: according to Reid, we only experience what is going on right now. Given that change cannot occur at a single moment of time (or, more precisely, that we cannot experience change as occurring at a single moment of time), we can only be said to experience it in the "loose sense" with the help of memory. Simple accounts of such a *memory theory* argue that we are aware of change by perceiving what happens now, while remembering at the same time what has happened just before.[3]

In general, memory theories face a number of serious objections. One common objection is to point out that the phenomenology of memories is rather distinct from

---

[2]Realism is defended among others by Dainton (2000, 2008a, b, 2010, 2011), Foster (1982), Le Poidevin (2004, 2007), Phillips (2010), Tye (2003) and Zahavi (2007). Antirealism is defended among others by Dennett (1991), Kelly (2005), Mabbott (1951, 1955) and Plumer (1985).

[3]Philosophers that have defended memory based accounts of temporal perception include Le Poidevin (2007), Mellor (1998) and perhaps Phillips (2010).

that of perceptual experiences (cf. Dainton 2000; Kelly 2005). If our awareness of change was based on memory rather than on perceptual experience alone, then the phenomenology of change experience (or any temporal experience) would be rather different from genuine perceptual experience, but this does not seem to be the case. Memory theorists have a variety of answers to this (and other) objection(s) but I won't have time to go further into that here. Whether the objections against the memory theory prove devastating or not, the memory theory in general appears to have a problem in justifying that we do not perceptually experience temporal phenomena although it seems very natural and obvious to think that we do.

3 Specious Present Theories

All major realist accounts of temporal perception are *'specious present theories'*. The term 'specious present' was coined by the psychologist Clay (1882) but the theory of the specious present was made famous by William James (1890).[4] In contrast to others before him, James believed that we have (purely) perceptual experiences of change and succession. Rather than just making us aware of what is right here now, James thought that our experiences would make us aware of what happens over short intervals of which we are aware of *as present*. He called the period of time that is experienced as present, the specious present. The term 'specious present' adverts to the fact that the experienced present (the present as experienced) deviates from the objectively present time: whereas the objective present is standardly taken to be a moment, the specious present is supposed to have temporal depth. James referred to the specious present as a *duration-block* (Dainton 2010), which he characterized as follows:

> We do not first feel one end and then feel the other after it, and from the perception of the succession infer an interval of time between, but we seem to feel the interval of time as a whole, with its two ends embedded in it (James 1890: 610).

Thus, according to specious present theories in general, changes or successions are represented *all together as a whole* during the short interval that is perceptually represented *as present*. Although we are supposed to be aware of successions and changes 'all at once', they are represented as temporally extended—a somewhat puzzling claim that different specious present theories explain in different ways. I shall come back to this later. There are two major types of specious present theories, which I shall call, following Dainton, the *retentional theory* and the *extensional theory* (Dainton 2011).

[4]Although Clay was the first to coin the term 'specious present', it is relatively hard to retrieve more information about him or his work. According to Andersen and Grush (2009), the name 'E. R. Clay' is a pseudonym for the psychologist Robert Kelly. The book attributed to him, 'The Alternative: A study in psychology', was anonymously published in 1882.

4 The Retentional Theory of Temporal Perception

Traditionally, specious present theories all used to be retentional theories of temporal perception.[5] Like all specious present theories, retentionalists argue that we can perceptually experience *more* than what is momentarily present. The basis of the retentional theory is formed by a combination of three principles:

(i) Our perceptual experience is not confined to what is (was) momentarily present.
(ii) In order to have a perceptual experience of a temporally extended structure as temporally extended, all parts of that structure have to be experienced together.
And a principle that Miller calls
(iii) *The Principle of Simultaneous Awareness* (PSA): If we are directly aware of the immediate past, this awareness is located in the present. (Miller 1984: 109)

When we combine (i)–(iii), then the view is that to perceptually represent change, we need to be aware of all parts of the temporally extended structure at the momentary present. PSA is the doctrine that differentiates the retentional theory from the other realist account, the extensional theory (I will come back to that later). According to PSA, we are, *at any instant*, aware of what occurs over an interval of time. This distinguishes a succession of experiences from an experience of succession: we experience succession only when all parts of the succession are experienced together, in one instantaneous experiential act. It follows that, according to the retentional theory, one's experience needs to 'literally embrace' (Phillips 2010: 6) at a moment the entire temporal structure that constitutes change:

Diagram 1

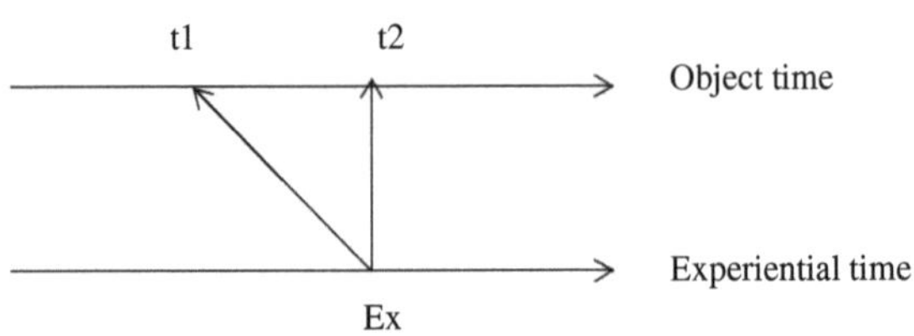

Consider the diagram: *at* t_2, the experience 'Ex' is a representation of what occurs over the objective time interval $t1$-$t2$, where $t1$ is in the immediate past and

[5]We find various versions of the retentional theory in Husserl (1905, 1964), Brentano (Brentano 1988), Meinong (1978 (1899)), Russell (1915) and Broad (1923, 1938). Husserl for example thought that to experience succession, we need to perceptually represent what is currently present and *retain* the immediate past, where what is retained (the *retention*) differs from memories in that they are sensory, and from perceptual experience in that what is represented is represented as past. Similarly, Brentano thought that we would 'intend' to the past parts of the succession as (recently) past. Meinong held that representations of the immediate past are 'reproductions' of perceptual experiences and Russell took them to be fading 'akoluthic sensations'. In all cases the representation of the past is distinguished from a memory. Broad, finally, held that we can actually *perceive* what pertains to a short interval stretching from the present into the past.

*t*2 is present. The interval *t*1-*t*2 is represented at *t*2 as present and thus constitutes the experienced or specious present.

Thus, according to the retentional theory, we perceptually represent at a moment things that have duration, and we represent them as having duration.[6]

This strikes many as counterintuitive. What seems to puzzle many is the idea that experiencing succession necessarily takes time, so it seems to be straightforwardly incoherent to assume that we experience succession at a time. However, it should be considered that temporal perception is unique in that experiences not only represent objects as having temporal properties, they themselves instantiate temporal properties. When an experience represents the duration of some thing, the experience, that is, the *experiential act* itself, has a certain temporal shape—it may last for some time or occur at a moment, for example. It is by no means clear why the temporal structure of the experiential act could not come apart from the temporal structure of what is represented, as illustrated in diagram (1).

5 The Extensional Theory of Temporal Perception

The extensional theory of temporal perception holds that we can perceptually experience temporally extended structures such as changes and successions.[7] It rejects the idea that all parts of a succession are represented at a single moment, while adhering to the principle that successions are represented 'together, as a whole'. Consider two successive tones, Do and Re. The extensional theory holds that we can perceptually experience the succession of (Do-Re) and that the perceptual act is concurrent with what is experienced: if the succession Do Re seems to occur over two seconds, say, then the experiential act will also take two seconds. During this short period, the extensionalist's specious present, we are perceptually aware of (Do-Re)—both tones seem present to us, although as occurring in succession.[8] This appears initially counterintuitive: as Le Poidevin (2007: 87) asks, how can something that is experienced as present not be experienced as simultaneous? The extensionalist answer is that Do and Re are experienced as *phenomenally* present (rather than metaphysically present), in the sense that both have a psychological 'presence', although they appear in succession. How does the

[6]Traditional accounts of the retentional theory have not formulated in terms of representationalism (Broad for example was a sense-datum theorist). In contemporary versions however, representationalist accounts have been given as well (Tye 2003; Kiverstein 2010). In presenting the theory, I stick to a minimal representational framework.

[7]Extensional accounts of temporal perception have been early supported by Stern (1897/2005) and Mundle (1954, 1966), and lately defended in the more elaborate form of the *Overlap Theory* by Foster (1982) and Dainton (2000, 2001, 2003, 2008a, b).

[8]It is notoriously difficult for extensionalists (and retentionalists) to determine the duration of a specious present. Dainton for example tentatively estimates it to last for 'half a second or less' (Dainton 2000: 171).

extensional theory differentiate between a succession of experiences and an experience of succession? This is accounted for by the fact that temporally extended experiences are *phenomenally unified.* A phenomenally unified experience (let us call it 'total experience') is phenomenally unified if *at any time t* during the experience the subject is phenomenally aware of *all* of the temporal parts of the experience, more exactly, of all temporal parts including those occurring at other times than t. We are able to perceptually represent what happens over a period of time, by virtue of the fact that successive partial experiences are phenomenally united during the concurrent interval that is the specious (psychological) present.

Diagram 2

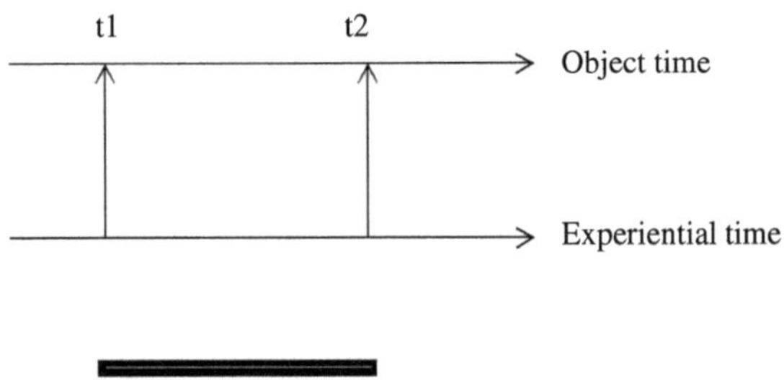

Total experience

The extensional specious present, depicted by the bar in the diagram, refers to the real-time interval in consciousness over which a total experience occurs. During that interval everything is represented *as* present.

A notable weakness of the theory concerns the explanation of experienced order. The problem is that the relation of phenomenal unity is symmetrical and thus cannot account for the asymmetry of direction in which (partial) experiences succeed each other. Dainton explains the directedness and the order of experiences with the fact that each partial momentary experience instantiates irreducible and basic temporal properties, properties of "flow", thereby possessing an 'inherent direction' (ibid.). The appeal to primitive and temporal properties leaves many unsatisfied. It leaves a taste of "*ad hocness*" which might be a disappointment for otherwise well-disposed readers of the extensionalist doctrine.

Here ends my very brief exposition of the current main theories of temporal perception in contemporary analytic philosophy. Many important themes have been left out—the relation between metaphysical theories of time and theories of temporal consciousness, the metaphysical status of the specious present, or whether the competing explanations tell us different stories about how we deal with time in everyday practical life. These topics are simply too big for a very brief introduction. My hope is that I and the other authors have contributed in providing the reader with a first insight into a very complex and fascinating topic.

References

Andersen, H., & Rick, G. (2009). A brief history of time-consciousness: Historical precursors to james and husserl. *Journal of the History of Philosophy, 47*(2), 277–307.

Brentano, F. (1988). *Philosophical investigations on space, time and the continuum*. Translated by B. Smith. Beckenham: Croom Helm

Broad, C. D. (1923). *Scientific thought*. London: Kegan Paul.

Broad, C. D. (1938). *An examination of mctaggart's philosophy* (Vol. II). Cambridge: Cambridge University Press.

Clay, E.R. (1882). *The alternative: A study in psychology*.

Dainton, B. (2008a). The experience of time and change. *Philosophy Compass, 3*(4), 619–638.

Dainton, B. (2008b). Sensing change. *Philosophical Issues, 18*(1), 362–384.

Dainton, B. (2010). Temporal Consciousness. In E. N. Zalta (Ed.), *The Stanford encyclopedia of philosophy*.

Dainton, B. (2000). *Stream of consciousness: unity and continuity in conscious experience*. Routledge.

Dainton, B. (2001). *Time and space*. Ithaca: McGill-Queen's University Press.

Dainton, B. (2003). Time in experience: Reply to Gallagher. Psyche, http://psyche.cs.monash.edu.au/symposia/dainton/gallagher/r.pdf.

Dainton, B. (2011). Time, passage and immediate experience. In Craig Callender (Ed.), *Oxford handbook of philosophy of time*. Oxford: Oxford University Press.

Dennett, D. C. (1991). *Consciousness explained*. Penguin.

Foster, J. (1982). *The case for idealism*.

Husserl, E. (1905, 1964). *The phenomenology of internal time-consciousness*, M. Heidegger (Ed.). Bloomington: Indiana University Press.

James, W. (1890). *The principles of psychology*. New York: Dover.

Kelly, S. D. (2005). The puzzle of temporal experience. In Andrew Brook (Ed.), *Cognition and the brain: The philosophy and neuroscience movement*. Cambridge: Cambridge University Press.

Kiverstein, J. (2010). Making sense of phenomenal unity: An intentionalist account of temporal experience. *Royal Institute of Philosophy Supplement, 85*(67), 155–181.

Le Poidevin, R. (2004). A puzzle concerning time perception. *Synthese, 142*(1), 109–142.

Le Poidevin, R. (2007). *The images of time: An essay on temporal representation*. Oxford: Oxford University Press.

Mabbott, J. D. (1951). Our direct experience of time. *Mind, 60*(April), 153–167.

Mabbott, J. D. (1955). The specious present. *Mind, 64*(July), 376–383.

Meinong, A. (1978 (1899)). On objects of higher-order and their relationship to internal perception. In M. L. Schubert Kalsi (Ed.), *Alexius meinong*, (pp. 137–208). The Hague: Martinus Nijhoff.

Mellor, D. H. (1998). *Real time II*, Routledge.

Miller, I. (1984). *Husserl, perception and temporal awareness*. Cambridge, MA: MIT Press.

Mundle, C. W. K. (1966). Augustine's pervasive error concerning time. *Philosophy, 41*(156), 165–168.

Phillips, I. (2010). Perceiving temporal properties. *European Journal of Philosophy, 18*(2), 176–202.

Plumer, G. (1985). The myth of the specious present. *Mind, 94*(January), 19–35.

Russell, B. (1915). On the experience of time. *The Monist, 25*(2), 212–233.

Stern, W. L. (1897, 2005). Mental presence-time. In C. Wolfe (Ed.), *The New Year book for phenomenology and phenomenological research*. London: College Publications.

Tye, M. (2003). *Consciousness and persons: Unity and identity*. MIT Press.

Zahavi, D. (2007). Perception of duration presupposes duration of perception—or does it? Husserl and Dainton on time. *International Journal of Philosophical Studies, 15*(3), 453–471.

Part II
Language and Thinking of Time

Introduction

Roberto Evangelista

Abstract This chapter deals with the link between language and epistemology of time, according with the paper issued in this session. The arise of natural science, in Early Modern Age, makes possible the constitution of an epistemic paradigm which connects himself to the definition of human nature, and allows to research objective rules for the human history and human language. The Early Twentieth century, breaks (or tries to break) this standard, as is possible to see in the essays proposed in this part of the book.

The epistemological problem can be synthesized by two brief questions: what is science (and therefore knowledge) and what can we know? The relationship between science and knowledge takes particular shape in modernity, precisely, when we attempt to overcome the Aristotelian schema. One certainty is that the break with Aristotelian reasoning has been neither abrupt nor definitive, yet scientific methods have changed radically, especially in terms of considering the subject as the protagonist of knowledge. Man: the protagonist of the relationship between subject and object, the *holder* of ideas, the centre of self-awareness of the actions he performs. Although he is also a particular expression of natural force and thus intensely bound to nature, he is at the same time the creator of technical, political and cultural instruments and strategies.

Knowledge of nature becomes a question of *measure,* or rather of proportion between the observation of the subject and the activity of the object. Attempts to measure and formalize observation are more precisely a means of communicating the language of nature (which according to Galileo is a language written in mathematical letters). However, the observation and measurement of nature does not construct a *realistic* idea of science and epistemology. This is simply because such an idea does not exist; the concept of nature which is not interpreted culturally, nor subject to historical change, is pure mythology. It is no coincidence that the measurement of nature, albeit by possibly tortuous but decisive pathways, soon

R. Evangelista (✉)
National Research Council, Institute for the History of Modern Philosophical and Scientific Thought, via Porta di Massa 1, 80133 Naples, Italy
e-mail: evangelista.roberto@gmail.com

F. Santoianni (ed.), *The Concept of Time in Early Twentieth-Century Philosophy*, Studies in Applied Philosophy, Epistemology and Rational Ethics 24,
DOI 10.1007/978-3-319-24895-0_13

reveals the history of the natural world. In this way we find the objective content of time, or rather its *object*. The question *what is time* acquires a new meaning, because it becomes *of what does time consist*. The perspectives exposed by this detour would appear unforeseeable.

Time consists of natural mutations: gases, disintegration, earthquakes, sand, mud, decay and rebirth. The human body changes as nature changes around the body, and the mind changes with it. Above all however, the mind of man finds different ways to express his perceptions of these natural mutations, to explain their causes and predict their outcomes. The measurement of nature becomes a prediction of natural events. This action has both the characteristics of prophecy and of scientific reasoning. Undoubtedly, our attention turns to Vico, in whom this schema has revealed its most rewarding outcomes and its boundaries to our understanding of nature. But not only this.

Everything becomes measurable: Galileo and Descartes (the latter with his theory of neural transmission) allow us to formalize many if not all aspects of the human being. Or at least they let us think that this is possible. Even the relationship between the individual, the constitution and the maintenance of political society become elements of nature, which respond to necessary laws and rules. Vico's standpoint is a long way from here, and yet this is the premise on which man loses his absurd centrality of *imperium in imperio* (as Spinoza writes) and is brought back to his natural element, the earth. Although the history of the world and the history of mankind do not coincide, they influence each other profoundly and reference points of the old order are broken. The law of nature is amoral, it presents no ending and depicts a world in which every event may be justified. Discovering that the nature of man is amoral, when it is not primitive, focuses scientific observation on the problem of power. It is a hard task to govern human nature, when it is impossible to describe a human being who does not consider his most basic needs.

There are various answers to these complexities. Reading Hobbes and Locke, one can see how the answer to the anarchic nature of man lies in an artificial product: language. All philosophers are in agreement when language has no transcendent references to human action but only references related to experience, because these definitions are the products of convention. Certainly, Hobbes is an extremist on this point: language functions equally for everyone, only if another yet more artificial product is added. The Leviathan is the artifice which allows authority to exercise its defining power.

Political science allows us to delimit space for the rulings on the language, civil customs, doctrines and scientific beliefs. It is of no importance that this is made possible by mechanisms which are more or less authoritarian: the process that modernity designs proceeds from science to power and vice versa. The attempt is to govern that which is known as ungovernable: nature and man. To govern human nature means to be aware of his/her most unpredictable aspects. Memory, imagination and passion become more pressing in the philosophical framework and constitute its most important aspect. Imagination and memory are an entrance through which time becomes one of the ways in which to exercise power. If earthly time is ungovernable, that of man can be managed through the memories and desires of each individual.

Of what, then, does time consist? Of past experiences and of future plans. The object of time becomes all that we are capable of knowing and all that is useful in terms of predictability of the future. The formation (the history) of language, as well as the shared memories and the traditions of a population are the direct consequences of scientific observation. The same can be said of plans and the quest for utility; prediction and predictability originate from scientific knowledge. This may be one of the principal aspects inherited from Cartesian philosophy. It is a coherent science, a view which organizes multiplicity, allowing us to see the human world as an ordered entity, in which before and after acquire a meaning in the present.

And yet this viewpoint may fragment. The unity of knowledge and the unity of history do not resolve the conflict between the individual and society, between culture and nature, between power and authority. However, only if we take this unity as given, can conflict emerge in all its force. The relationship between time and epistemology and between time and knowledge have not cancelled this typical Early Modern conflict. If anything it has complicated the conflict by introducing new players, nevertheless, the question still remains as that of access to knowledge and the answer in each case is that knowledge is mediated by time.

Each one of us experiences ourselves *in time;* otherwise, as Vico writes in *De antiquissima* criticizing Descartes, we would have awareness but not knowledge. In contrast to the so-called vulgar materialism, in which time and space are objective elements, it would appear that the idea of time is slowly affirmed as an *auxilium imaginationis,* as Spinoza declares. Imagination is however a form of knowledge, a way to confront and construe reality and one's own identity. The first self-consciousness is imaginary, as are personal and cultural identity. In contrast, time also becomes via *negationis* the opportunity to know *sub specie aeternitatis.* Is it not possible that the eternity of which Spinoza speaks is in fact time emptied of its passing? Could it not be the container of all changes and all known times emptied of its contents? Spinoza's eternity could therefore be read as an attempt (perhaps the last) to give time an ontological dimension, in which time becomes eternity and science becomes the science of eternal things.

The reasoning of the Twentieth century touches on these arguments without necessarily seeking solutions; on the contrary it often provokes contradictions. The subconscious confrontation between society, nature and power has guided the topic of human identity. Inaugurated by Freud and reclaimed by Freudianism, it can be found in the reflections of Lacan, Foucault and in particular Deleuze, but even previously, in phenomenological psychology and in the mediations of the Hermeneutical circle as well.

Inclusions in this section deal with the question of time and epistemology, and of the solutions and opinions of authors who have returned to this question. A central issue however remains: the difficulty of the relationship between subject and reality. This relationship is problematic not only because of the uncertainty of human knowledge, but above all because of the uncertainty of communication. The relationship between time and epistemology since its first premises, and particularly in light of its contemporary results, has gone beyond the concept of time as mere quantity and constitutes it as a measure of the knowledge we have of ourselves and the world. In this context, as revealed by the contributions of this section, language is the measure of how knowledge may be expressed, which is acquired through time.

Giorgio Rizzo, in his essay on Wittgenstein, emphasizes a distinction between the time of experience and existence and quantifiable time as a measure. Nevertheless, the problem is still complicated by the relationship between time and linguistic expression.

The proposal of Heidegger is considered in the work of *Simona Venezia*. Here, we find the relationship between time and understanding, which defines the Hermeneutical nucleus of reflection by the author of *Sein und Zeit*. The essential condition of man can only be understood if we abandon the traditional conceptions of metaphysics. The connection between epistemology and time is thus directed towards the possibility of comprehending and interpreting the existential state of man.

The contribution of *Fabrizio Lomonaco* concentrates on the critical position of Cassirer and of the attempt to know a form *a priori*, to understand the different aspects of the spirit. Symbolic forms are forms of thought which live in an eternal and immanent present.

Giuseppe Cacciatore, in his text dedicated to Ricoeur, approaches the question from the link between time and existence. Of particular importance are the anti-metaphysical reflections of Ricoeur, in which the connection between history and narration is emphasized. The plurality of history is the way the link between temporality and epistemology can be inserted in an ethical dimension.

The essay of *Agostino Cera* covers a similar terrain. In fact, Foucault's perspective is analyzed according to the existence of two different historical phenomenologies: that of epistemic time and that of evenemential time. If the first temporality constitutes historical time taken when it actually happened, the second temporality describes the time of knowledge. Both of them, however, serve to delimit the space of events and the knowledge of the subject.

Lacan's perspective, on the contrary, replaces time at the centre, as observed by *Marco Castagna*, and restores its centrality in the reconstruction of the history of the subject. The objective of psychoanalysis, however is not simply memory but also the rewriting of a personal history. Language therefore returns as a protagonist and its profound relationship to a time once lived re-emerges, consisting of memories and plans.

The framework of the relationship between epistemology and temporality, which we have tried to trace from its beginnings, thus appears even more complex. Twentieth century philosophy introduces new angles and perspectives, leaving this question unresolved but providing a new centrality. This has been demonstrated precisely and clearly by this section.

Language and Thinking of Time Maps

Flavia Santoianni

1 Time and Epistemology Atlas El Map

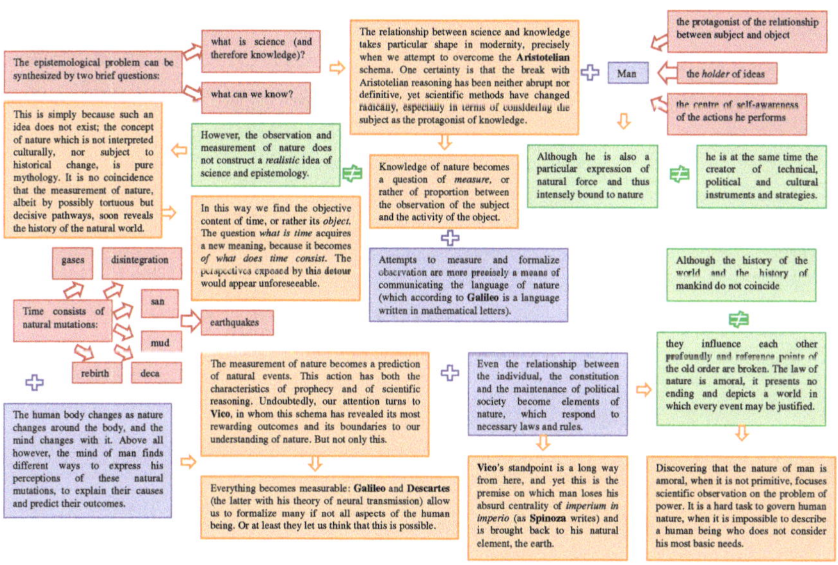

Time and Epistemology – EL map – Flavia Santoianni

F. Santoianni (✉)
Philosophy Section, Department of Humanities, University of Naples Federico II, via Porta di Massa 1, 80133 Naples, Italy
e-mail: bes@unina.it

F. Santoianni (ed.), *The Concept of Time in Early Twentieth-Century Philosophy*,
Studies in Applied Philosophy, Epistemology and Rational Ethics 24,
DOI 10.1007/978-3-319-24895-0_14

2 Time and Language Atlas El Map

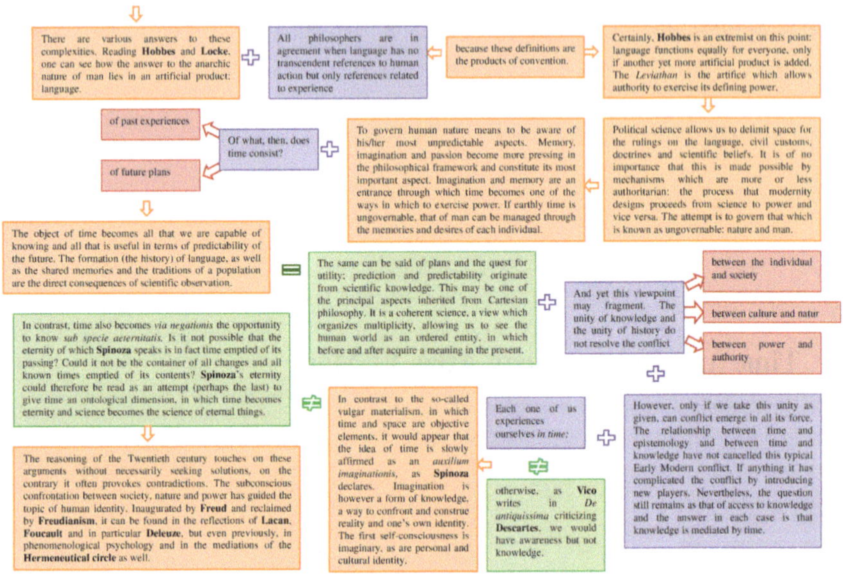

Time and Language – EL map – Flavia Santoianni

3 Cassirer, Wittgenstein and Heidegger Atlas El Map

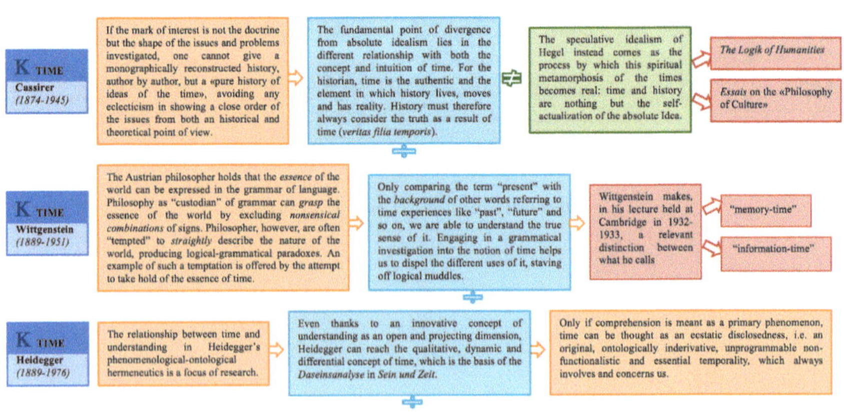

Time in Cassirer, Wittgenstein, and Heidegger – EL map – Flavia Santoianni

4 Lacan, Ricoeur and Foucault Atlas El Map

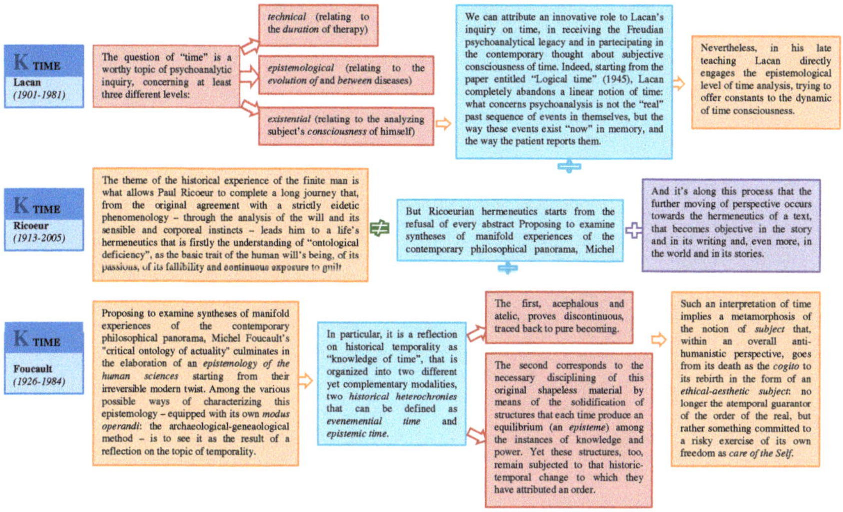

Time in Lacan, Ricoeur, and Foucault – EL map – Flavia Santoianni

5 Atlas Map

Atlas map – *Philosophers are located near their birthplaces. The Geographical Boundaries of Countries May Differ in Comparison with the European Geography of the Early Twentieth-Century*

Historiographical Language and Temporality in Ernst Cassirer

Fabrizio Lomonaco

Abstract This chapter is dedicated to the Cassirer's theories concerning *The Logik of Humanities* and others Cassirer's *Essais* on the «Philosophy of Culture.» The first part presents the concrete passages of mutual criticism on language and its relations with art and in relation to the interpretation of the modern history of philosophy. Brief final remarks recapitulate some radically different ways of understanding idealism with respect to the spiritual activity of man in the history.

This chapter is dedicated to the Cassirer's theories concerning *The Logik of Humanities* and others Cassirer's *Essais* on the «Philosophy of Culture.» The first part presents the concrete passages of mutual criticism on language and its relations with art and in relation to the interpretation of the modern history of philosophy. Brief final remarks recapitulate some radically different ways of understanding idealism with respect to the spiritual activity of man in the history.

Ernst Cassirer was born in 1874 in Breslau; in 1886, his family moved to Berlin, where he attended the lectures of Paulsen first, and then of Simmel, on Kant. In Marburg, he studied philosophy and took the courses given by Cohen and Natorp. In 1898, he gave a first draft of the *Leibniz System*; in 1902, he was in Berlin with his wife, Toni Cassirer. In 1919, he obtained the chair at the University of Hamburg and in 1929, he became Rector; in 1933, with the coming to power of Hitler, he left Germany on account of his being Jewish, and until 1935, he taught in Oxford. He moved to the University of Gothenburg for a few years and afterward, in 1941, he moved to the United States, where he taught as a visiting professor at the University of Yale.

Among his works we should recall: *The Problem of Knowledge: Philosophy, Science, and History since Hegel* (in four volumes, 1906–1920), *Substance and Function* (1910), *Kant's Life and Thought* (1918), *Individual and the Cosmos in the Renaissance Philosophy* (1927), *The Philosophy of the Symbolic Forms* (in three

F. Lomonaco (✉)
University of Naples Federico II, Department of Humanities, Philosophy Section,
via Porta di Massa 1, 80133 Naples, Italy
e-mail: flomonac@unina.it

F. Santoianni (ed.), *The Concept of Time in Early Twentieth-Century Philosophy*,
Studies in Applied Philosophy, Epistemology and Rational Ethics 24,
DOI 10.1007/978-3-319-24895-0_15

volumes, 1923–1929), *The Philosophy of the Enlightenment* (1932), the *Essay on Man* (1944). He died on April 13, 1945, in Princeton.

In a speech held in Gothenburg in 1935, the task entrusted to philosophy betrays the very original *style* of the Cassirerian historiography, already presents in the reconstruction of the thought of the Enlightenment of 1932:

> The historian, philologist, linguist, ethnologist, and the investigator of myth and history of religion are involved with the forms of culture. But here philosophy (…) starting with such forms (it) must go back and ask about the formative powers, about the type of spiritual functions and energies which have produced and made possible these shapes of the human spirit (Cassirer 1935: 56, 65).[1]

The project of a history of the «modern spirit» involves three well-known monographs published between 1927 and 1932, respectively, *Individual and Cosmos in the Renaissance Philosophy*, *The Platonic Renaissance in England and the Cambridge School,* and *The Philosophy of the Enlightenment*. In the introduction to this last one, the author emphasizes the rigorously unitarian reason of the methodological intention, based on the search for «the inner formative forces» working from the inside, a «phenomenology of the philosophic spirit,» caught and pushed throughout in its fundamental *character* and its *mission* (Cassirer 2009^2: XII).[2] To this end, the interpreter does not propose a detailed historical account of Enlightenment thought. His purpose is to isolate the center of *gravity* in a complex spiritual becoming, in order to procure not a quantity of doctrines but the dominant «new form of thought» in the unity of its origin and development. If the mark of interest of the renewed investigation is not the doctrine but the *shape* of the issues and problems investigated, one cannot give a monographically reconstructed history, author by author, but a «pure history of ideas of the time», avoiding any eclecticism in showing a close order of the issues from both an historical and theoretical point of view:

> (…) It is insufficient to remain satisfied with the simple existence of facts as they present themselves in morals and customs, in political structure and social behavior, in codes of law and articles of religious faith. Instead of being accepted on faith their origin must be discovered and their cause proved valid by the criteria of reason, otherwise the existing institution can expect no recognition (Cassirer 1931: 401; Cassirer 1955^2: 345–346).[3]

All the facts must be reported to the corresponding creative energies, allowing us to continually recreate those contents that would otherwise be fixed and locked in their objective determination. The historiographical work attests that the modern formation of objectivity lays in tracing the spiritual conception that produced it, in the concrete historical development. Of an entire age, we must see the continuous

[1]See also Ferrari (1986): 102. Martirano (1997): 396–408. On "The Marburg School" see also Skidelsky (2008): 22–51.

[2]For more extensive critical literature in this and in the following notes I refer to Raio (1991): 213–246.

[3]See Lomonaco (2011). The other passages translated into English were compared with the edition of the *Gesammelte Werke* (Hamburg: Felix Meiner Verlag) and with that Italian cited below.

and never-dogmatic 'becoming,' whose goal is not a defined product, but the constitution of sense of another subject, that may capture and develop the spiritual behavior of what the subject has received. This is where the awareness of the multidimensional nature of both philosophy and philosophical historiography comes in opposition to the resolution of history and philosophy in the *contemplation*—and progressive affirmation—of a philosophy, embraced in a final form only. The true meaning of every era in history does not depend on the simple development of the themes of its related past culture but by the spontaneity of a spirit that regenerates those themes, in this way discovering its most authentic and deep being, freeing its energy and reaching its most characteristic «tensions and solutions.» The guarantee of novelty in all areas lays in the use of a new method, the event of a theoretical turn of great significance. The unity of philosophy does not arise—as in metaphysics—by the self-development of the Hegelian *Idea*, but as a postulate of the reason that, according to a known expression of Goethe, switches the unity of history from metaphyisical to methodical (Cassirer 2009[2]: XI; Cassirer 1978, I/1: 33, 34).[4] In this breakthrough, there is a positive significance to the development of the Kantian lesson of the masters of Marburg in the direction of a greater integration of the humanities with the natural sciences, which occurred in the sign of Dilthey and his anti-Hegel criticism to the romantic condemnation of the Enlightenment.

In the inner contrast with the Romantic bias against Enlightenment, it was necessary to understand the efforts of the scientific emancipation and the safeguarding of the action of the reason in all different fields of the spirit. The problem of the *Erkenntnisproblem* (1920) had moved from the ground of a history of the problem of knowledge to the consideration of the gradual emergence of a spiritual autonomy of man. From the Renaissance to post-Kantian idealism, the question was to follow the claim of the free activity of man without falling into a metaphysical detour. In this theoretical and historiographical intention, we find the most general conflict between *critical* idealism and *absolute* idealism that Cassirer, in the concluding paragraph of the chapter on Hegel had made evident, pointing out how the spirit, in the philosophy of the absolute idealism is never past nor future, but present time, and that there is always the possibility that a particular present belies the immanent eternal present. The whole becoming of the idea is concentrated in «one supreme point» that empties of independent meaning «the whole previous development» (Cassirer 1978: 463–464[5]). Commenting on a well-known Hegelian page, Cassirer points out a difference between the kind of remembrance which is reached through empirical knowledge, in the history of art of mythology or religion, and the one generated by metaphysical knowledge. The fundamental point of divergence lies in the different relationship with both the concept and intuition of time. For the historian, time is the authentic and—in a sense—the only dimension of his thought; it is the element in which history lives, moves, and has reality.

[4]On this point see Ricci (2009): 469, 470.

[5]See Martirano 1997: 411 ff.

History must therefore always consider the truth as a result of time (*veritas filia temporis*). But *speculative* philosophy cannot accept this approach: even when you are interested in the phenomena of time, the ebb and flow of history, this sphere proves to be unsatisfying, it tends to *contemplate* the realm of reality *sub specie æternitatis*. The speculative idealism of Hegel comes as the process by which this spiritual metamorphosis of the times becomes real: time and history are nothing but the self-actualization of the absolute Idea. Taken by itself, the idea must be free from all constraints and determinations of time, with no past, and no future: it is absolute and omnipresent. And philosophy holds together art and religion, unifying them in a single spiritual vision, raising them, thus, to the sphere of self-conscious thought. Through this, Hegel believes to have realized the true reconciliation of freedom and necessity. The way in which the absolute spirit comes to itself is a necessary way; the arrival point where this path ends is self-consciousness, that is to say the absolute freedom of the spirit.[6]

In an essay of 1936 Cassirer goes back to the marks of 'critical' idealism, warning that it imposes a different task, because it does not pretend to understand the content and goal of culture in the light of a logical deduction of every single moment in time nor does it intends to offer a metaphysical description of the universal plan of development. It does not judge that the single stages and processes by which the universe of culture is built lack effective and real unity, nor that they are nothing but scattered fragments, individual expressions of the human mind in a plurality of different directions. Despite their differences, they possess an intrinsic unity that cannot be defined or explained in metaphysical terms or described in formulas of mere substantiality, nor in the frame of an historical, natural, and fatalistic system. From these pages, a different relationship between history and philosophy emerges in relation to the intuition of time and the concept of idea that Cassirer defines as a pure and infinite task, which needs to be in consonance with critical idealism that has Kant as a source, and that common world to whom every individual conscience belongs. Having shown, therefore, the archaic character of the positivistic naturalism, it was matter of supporting—with the Kantian Copernican revolution—a general criterion of unification of knowledge: not so much a unification of *contents,* but a unification of *meanings*, understood and defined in functional terms, i.e., in terms of relationships, transactions, and action. And this, because it is not something given, but an ideal idea that has to be understood dynamically. It must be "produced" and in the conditions of this production lays the meaning and the ethical value of culture. *Critical* Idealism describes in detail the various forms of it, those of nature and science, religion, history, and art. No one can escape the *medium* of the form, as Cassirer warns, stepping away from the neocriticists, Cohen and Natorp, who had missed the sense of the multidimensionality of the logic of the structural spiritual principle that is realized in all the different forms of culture. Of those mentioned forms, the order sought does not stop at a universal formula expressing the absolute nature of the

[6]See Rovatti 1968.

spirit and the necessary sequence of its individual phenomena, nor does it claims to describe or predict the future course of the history of culture. The desire is to be able to come to a sort of grammar and syntax of the human spirit, for a review of its various *forms*, to get an *idea* of the general rules that govern them. These rules do not obey a predetermined pattern, which can be described once and for all on the basis of an aprioristic proceeding. All one can do is to follow the slow development that occurs in the history of the various forms and eventually indicate the *milestones* of it. To find them, one has to address the special sciences, accept the data they provide, by analyzing the nature of the different *functions* from which the phenomena, taken together, depend. That unity that Cassirer calls the «unity of symbolic thinking» cannot be abstracted from its various manifestations; it cannot be conceived as a separate existence. The main purpose of all forms of culture is to build a common world of thought and feeling; a human world that wants to be a common world participated in which «each individual consciousness» has to reconstruct itself «in its own way and by its own efforts» (Cassirer 1936, 1979: 90, 1981: 99).[7] The aim of the modern philosophy of culture is to keep together all the different forms and junctions of knowledge, founding a plurality of manifestation of the *kultur* on the possible «construction of a common world» with human action as a unifying center. In the development of the problem of the foundation of the logic of *Kulturwissenschaften*, philosophy of culture tries to evade «both the Scylla of naturalism and the Charybdis of metaphysics» (Cassirer 1979: 39–40, 46).[8] Inspired by the famous motto of Goethe (*im Anfang ist die Tat*), Cassirer's considerations of show that the different forms of culture are expressed in a common making in which the human subjects come together and understand each another, to the point that the symbolic forms can be defined as "the typical media created by man, and in virtue of them he can separate himself from the world and, in this separation, join it even more firmly" (Cassirer 1979: 22, 46, 70–71). This should explain the relationship between theory and historiography in the reconstruction of philosophical issues and problems, around some significant nucleus, bound to document the development of the sciences of culture, establishing the *systematic* formation of them. The study does not focus on neither the works of art nor on the products of any mythical or religious thought, but on the mental faculties needed in the production of such works. Only by managing to gain an insight into the nature of these faculties and understanding them in their structure, in order to conceive how they differ from each other despite their mutual cooperation, we will reach a *new* vision of the *character* of human culture. In this framework, it is necessary to move away from a formal logic where the concept is the result of an abstraction of all the peculiar features of the single objects, leaning, then, toward a model of *abstract universal concept*. The purpose of Cassirer's reflection in the Forties will be to highlight the contribution of a transcendental conception of logic able to turn the concept into a

[7]See Martirano 1997: 412 and note.

[8]On this theme see Lomonaco (2013).

genuine instrument for understanding the different areas of the spirit, a clear 'a priori' point of view constituting a particular understanding of the world. From this point of view, every symbolic form is a form of thought in which each cultural phenomenon is confined to the world of the "natural objectivity" and, as any other object, having "its own position in space and time" (Cassirer 1979: 39–40).

The task of founding the sciences of the spirit is at the heart of the problem *morphology*, bound to investigate not only those sciences themselves and their methods or generally enunciating their hallmarks, but also first of all the multiple forms of the human understanding of modern. The *philosophy of the symbolic forms* wants to be an execution of this general methodological design; it rests on the recognition of an original and specific formation function, not only for what concerns the understanding, but also in each other sphere of human experience. This recognition is needed to transform philosophy from being a *critique of reason* into a *philosophy of culture,* able to recognize some fundamental common traits—typical of the various formation activities (Cassirer 1936: 80–81 and ff., 1981: 89–80 and ff.).[9]

It is no coincidence that many years before the theory of the *animal symbolicum*, entrusted to the pages of the *Essay on Man*, the German scholar opened his philosophical horizon to a complex attempt of mediation—bound to mark the end of his production—between the transcendental foundation of the philosophy of culture and the "self-knowledge of man," that the crisis of the contemporary culture made more and more problematic. For all that, it was necessary not to lose the positive tension between condition and conditioned, in order to understand that it is not a question of tracing a line backward, going from man as a natural entity to the forms of culture, but on the contrary to understand that man is constitutively this same ability to live within the forms of culture, as attested by comparison with the themes of philosophy of life (Scheler) and anthropology (Plessner) in the unpublished works of 1928 *Zur Metaphysik der symbolischen Formen* (Cassirer 2003; Ferrari 1996: 318, 321–323).[10]

Culture cannot be explained and defined in terms of necessity, but in terms of freedom, and understood in an ethical sense rather than a metaphysical one. Coherent is the opposition to Simmel, who speaks of "tragedy of culture," placing a contrast between the life of the subject that creates value and the values themselves, objectified and made lifeless in this objectification, emphasizing the loss of subjective vitality in the forms of the objective spirit (Cassirer 1979: 99 ff.). For Cassirer, the risk of such a diagnosis is that it lead to a *mysticism*, whereas polarity, in the relationship between different subjects, mediated by the products of the culture that "is dialectic, that is dramatic" (Cassirer 1979: 102)[11] is entrusted to the ongoing working of men and of their spiritual action, unsure of the goal, but still unstoppable.

[9]Lomonaco 2012.

[10]cf. Randazzo 2005.

[11]See Simmel (1976). See Martirano (1990): 442.

References

Cassirer, E. (1931). Enlightenment. In Cassirer, Ernst. 2004. Aufsätze und kleine Schriften (1927–1931). Text und Anmerkungen von Tobias Berben [*Gesammelte Werke. Hamburger Ausgabe*, hrsg. von Birgit Recki, Band 17. Hamburg: Felix Meiner]; Italian translation: Cassirer, Ernst. 1955^2. *Dall'Umanesimo all'Illuminismo,* P. Oskar Kristeller (Ed.), Firenze: La Nuova Italia.

Cassirer, E. (1935). The concept of philosophy as a philosophical problem. In Cassirer, E. (1979). *Symbol, myth and culture. Essays and lectures of Ernst Cassirer 1935–1945*, D. P. Verene (Ed.), New Haven and London: Yale University Press; Italian translation: Cassirer, E. (1981). *Simbolo, mito e cultura*, D. P Verene (Ed.), Roma-Bari: Laterza.

Cassirer, E. (1936). Critical idealism as a philosophy of culture. In Cassirer, E. (1979). *Symbol, myth and culture. Essays and lectures of Ernst Cassirer 1935–1945*, D. P. Verene (Ed.), New Haven and London: Yale University Press.

Cassirer, E. (1961). *Philosophie der symbolischen formen*, E. Arnaud (Ed.), (vol. I). Firenze: La Nuova Italia.

Cassirer, E. (1978). *Das Erkenntnisproblem in der Philosophie und Wissenschaft der neueren Zeit*, A. Pasquinelli (Ed.), Torino: Einaudi.

Cassirer, E. (1979). *Zur Logik der Kulturwissenschaften*, M. Maggi (Ed.), Firenze: La Nuova Italia.

Cassirer, E. (2003). *Zur Metaphysik der symbolischen Formen*, G. Raio (Ed.), Firenze: Sansoni.

Cassirer, E. (2009^2). *The philosophy of the Enlightenment*. Trans. F. C. A. Koelln & J. P. Pettegrove. Princeton and Oxford: Princeton University Press.

Ferrari, M. (1986). Ernst Cassirer e la "Bibliothek Warburg". *Giornale critico della filosofia italiana, 65*, 91–130.

Ferrari, M. (1996). *Ernst Cassirer. Dalla scuola di Marburgo alla filosofia della cultura*. Firenze: Olschki.

Lomonaco, F. (2011). Carattere e stile nella storiografia filosofica del primo Novecento. Ernst Cassirer interprete dell'Illuminismo. In V. Cuomo (Ed.), *Carattere e stile* (pp. 23–41). Roma: Aracne.

Lomonaco, F. (2012). Ragione, storia e cultura ne "La filosofia delle forme simboliche". In F. Lomonaco (Ed.), *Simbolo e cultura. Ottant'anni dopo "La filosofia delle forme simboliche"* (pp. 17–40). Milano: F. Angeli.

Lomonaco, F. (2013). Per una logica delle scienze della cultura: Croce e Cassirer. *Bollettino Filosofico, 28*, 198–217.

Martirano, M. (1990). Scienze della cultura e storiografia filosofica in Ernst Cassirer. *Archivio di storia della cultura, 3*, 409–444.

Martirano, M. (1997). Ernst Cassirer e il problema dell'oggettività nelle scienze della cultura. In G. Cacciatore, G. Cantillo, & G. Lissa (Eds.), *Lo storicismo e la sua storia. Temi, problemi, prospettive* (pp. 396–408). Milano: Guerini.

Raio, G. (1991). *Introduzione a Cassirer*. Roma-Bari: Laterza.

Randazzo, I. (2005). Cassirer e l'antropologia filosofica. Introduzione a E. Cassirer, Introduzione alla storia dell'antropologia filosofica (Lezioni di Göteborg 1939–1940). *Archivio di storia della cultura, 18*, 387–421.

Ricci, S. (2009). Garin lettore di Cassirer. *Giornale critico della filosofia italiana, 88*, 457–477.

Rovatti, P. A. (1968). Funzione e significato della struttura temporale in Cassirer. *Il Pensiero, 13*, 118–146.

Simmel, G. (1976). Der Begriff und die Tragödie der Kultur (1912). In *Arte e civiltà* (pp. 83–109). Milano: Isedi.

Skidelsky, E. (2008). *Ernst Cassirer. The last Philosopher of Culture*. Princeton and Oxford: Princeton University Press.

Wittgenstein on Time: From the Living Present to the Clock Time

Giorgio Rizzo

Abstract Augustine's analysis of time in Book XI of *Confessions* represents for Ludwig Wittgenstein a good example of a philosophical question. In dealing with such theme, his thought undergoes relevant changes. In the *Philosophical Remarks*, written more than 10 years after the drafting of the *Tractatus*, the Austrian philosopher holds that the *essence* of the world can be expressed in the grammar of language. Philosophy as "custodian" of grammar can *grasp* the essence of the world by excluding *nonsensical combinations* of signs. Philosopher, however, are often "tempted" to *straightly* describe the nature of the world, producing logical-grammatical paradoxes. An example of such a temptation is offered by the attempt to take hold of the essence of time using propositions like "only the present experience has reality." The logical mistake hidden in this proposition lies in the *bad use* of the adjectival word "present" that would lose its *everyday* use and *functional* role in the language. Only comparing the term "present" with the *background* of other words referring to time experiences like "past," "future," and so on, we are able to understand the true sense of it. Engaging in a grammatical investigation into the notion of time helps us to dispel the different uses of it staving off logical muddles. Wittgenstein makes, in his lecture held at Cambridge in 1932–1933, a relevant distinction between what he calls "memory-time" and "information-time." If the first can be understood as a *now-centered system* mostly expressed by indexical sentences or as an *arrangement* relied on memory, and therefore inadequate to give any external physical criteria for time measurements, the second clearly refers to a *public chronology*, implemented by clocks, calendars, diaries, and so on. Grammatical misconceptions, however occur when we are "tyrannized" by a metaphor and not able to "move outside of" it. The Austrian philosopher makes no secret of preferring a characterization of time that rejects a truth-functional interpretation. As for the notion of "game" in the *Philosophical Investigations*, it is impossible to have something like a *common denominator* shared by every sentence involving time.

G. Rizzo (✉)
Department of Human Studies, University of Salento, via Stampacchia 45, 73100 Lecce, Italy
e-mail: giorgio.rizzo@unisalento.it

F. Santoianni (ed.), *The Concept of Time in Early Twentieth-Century Philosophy*,
Studies in Applied Philosophy, Epistemology and Rational Ethics 24,
DOI 10.1007/978-3-319-24895-0_16

Augustine's analysis of time in Book XI of the *Confessions* represents for Ludwig Wittgenstein a good example of a philosophical question (Malcolm 2001: 59–60). In dealing with such a theme, his thought undergoes relevant changes.

In the *Tractatus Logico-Philosophicus*, the Austrian philosopher declares that the *logical form* common to language and world cannot be *expressed* in language. Propositions, as a matter of fact, can only *show* the logical form of reality (Wittgenstein 1961: 4.121). In other words, we cannot express the *essence* of the world because the *sense* of it must lie outside it (Wittgenstein 1961: 6.41).

After all, Wittgenstein claims, the only propositions which have sense are those that refer to facts (*Sachverhalte*) "constructed by way of experiment" (Wittgenstein 1961: 4.031). The world is the *logical space* of the *contingent* (Wittgenstein 1961: 1.13), so that questions about the metaphysical order of the world like those concerning its temporal "fabric" are mere *nonsense*, out of order.

Such a strong position is lessened in the *Philosophical Remarks*, written more than 10 years after the drafting of the *Tractatus*, in which the essence, structure, or nature of the world find the opportunity to be expressed in the language. Note: not in singular propositions, but in the *rules* or *grammar* of language (Wittgenstein 1975: 85). Philosophy as "custodian" of grammar can *grasp* the essence of the world. By which means? Merely, by excluding *nonsensical combinations* of signs. Philosophers, however, are often *tempted* to describe the essence of world in straightforward fashion, producing in this regard logical-grammatical paradoxes (Kaspar and Schmidt 1992: 571).

One example of such a temptation is the attempt to take hold of the essence of time using propositions like "only the present experience has reality" (Wittgenstein 1975: 85). The logical mistake hidden in this proposition lies in the improper use of the adjective "present" that loses its *everyday* use (*Gebrauch*) in language in which it has a *functional* role. Only by comparing the term "present" with the *background* of other words referring to time experiences like "past," "future," and so on, are we able to understand the true sense of it. A proposition like that quoted above well represents the standpoint of the *solipsist* who asserts that only his experience has reality: "The proposition that only the present experience has reality appears to contain the last consequences of solipsism" (Wittgenstein 1975: 85).

The solipsist finds himself in the troubling situation of meaning *correctly* something that cannot be expressed (Wittgenstein 1961: 5.62) and this, paraphrased, would amount to saying that *solipsism of the present moment* is correct even if unspeakable (Hintikka 1996: 249), in keeping with what Bertrand Russell remarks about the construction of the world out of the *perspectives* of actual and potential perceivers in *Our Knowledge of the External World* (Russell 2009: 121). The nonsense of the propositions concerning the essential intimacy between the present moment and experience draws from the erasing of the *empirical-pragmatic* background which permeates such propositions in everyday usage. By giving them the grammatical status of *quasi-tautological* propositions, the philosopher makes every discussion about time or experience nonsensical. Assuming that "experience = (my) present experience," the temporal word "present" loses its *usefulness* in everyday linguistic commerce (Wittgenstein 1975: 85).

Putting experience and time on equal footing produces two other side effects, because if, on the one hand, time is "reduced" to eternity intended as "timelessness" (Wittgenstein 1961: 6.4311), on the other it is "mystified," that is, meant as a "queer thing," something that we cannot look into, while "all the facts that concern us lie open before us" (Wittgenstein 1958). If only we could scrutinize the grammatical abuse of the word time, "we shall feel that it is no less astounding that man should have conceived of it of a deity of time than it would be to conceive of a deity of negation or disjunction" (Wittgenstein 1958: 6).

Engaging in a grammatical investigation into the notion of time, though, helps us to dispel the different uses of it, staving off possible equivocations or logical muddles. As recorded by George Edwards Moore, the Austrian philosopher makes a relevant distinction between what he calls "memory-time" and "information-time," insofar as "in the former there is only earlier and later, not past and future" (Moore et al. 1993: 110).

If "information-time" clearly refers to a *public chronology*, implemented by clocks, calendars, diaries, and so on, memory-time would involve two different ideas:

(a) time understood as a *now-centered system* which comprises mostly *indexical* sentences;
(b) an arrangement that relies on memory, and therefore inadequate to give any external physical criteria for time measurements.

It is the second argument that couples time with the predicament of solipsism, that is, with *temporal solipsism*.

Moore notes:

> As regards Solipsism and Idealism he said that he himself had been often tempted to say 'All that is real is the experience of the present moment' or 'All that is certain is the experience of the present moment'; and that anyone who is tempted to hold Idealism or Solipsism knows the temptation to say 'The only reality is my present experience' (Moore et al. 1993: 102).

The opposition between "memory-time" and "information-time" hints at the one between *perspectival* and *public* modes of *identification*. "Memory-time," in particular, shows analogies with the way we refer to objects *via ostension*, making use of indexicals like "this" or "that." Perspectival reference is a paradigm case of reference to *phenomenological objects* so that, in the end, the dichotomy between "memory-time" and "information-time" would come to be that between a *phenomenological* and a *physicalistic* framework (Hintikka 1996: 243). Note that the difference between a public and a perspectival principle of identification implies the use of two different *cognitive systems*, even though this does not necessarily mean that we are endowed with two different kinds of knowledge or memory. To each mode of identity attribution pertains a particular *system of quantifiers* and "who," "what," "when" questions.

By way of illustration, if I ask "Who around here is Stephen Hawking," a scientist whose works I have read but whom I have not met, a satisfactory answer to such a question could be the following: "That one, in the corner."

For Wittgenstein, to treat indexicals like "this," "that" or even "I" as *names* entails conceptual puzzlement for such words refer only to "aspects" (*Aspekte*) of objects whose change in time raises the problem of finding a criteria of identification and recognition that is *independent* (from memory). For that reason, temporal objects and phenomenological ones share the same conceptual troubles.

The difference between a phenomenological and physicalistic conception of time becomes more evident when time is related to the *constitution* of the external world, that is, to being. What kind of world objects would import phenomenological time? The identity criteria constrained by such a framework would doubtless require "basic objects" (*individuals*) referred to as *time-slices* of persistent objects, like analogous Quinean "rabbit-stages." What about "information-time" ontology? In such a case, objects serve as *values* of quantifiers; they have to persist in physical time, for their identification criteria ask for *continuity in time*, that is, for the preservation *through time* of some essential attributes.

Taking heed of time not as a *fact of nature* or a *phenomenon* needing to be *penetrated*, one succeeds in the comprehension of its grammatical, not substantial, framework: "We feel as if we had to *penetrate* phenomena: our investigation, however, is directed not towards phenomena, but, as one might say, towards the '*possibilities*' of phenomena. [...]. Our investigation is therefore a grammatical one" (Wittgenstein 1953: §90).

Wittgenstein often resorts to analogies or *images* to prevent us from calling into play metaphysical propositions about time.

The frames in a film reel and the frames that are in front of the projector's lens on the screen embody, according to him, two different time concepts.

The frames in the film reel would affect a concept of time embracing past, present, and future moments, whereas, the image of time suggested by the frames on the screen would be that of the immediate experience, that is, time involving *memories* and *expectations* of a human agent. Time construed as *living present* cannot, for this reason, be separated from the *plans*, *purposes*, *obligations* which establish our "form of life" (*Lebensform*) and, in this respect, it entertains a *modal* (that is, not extensional) *structure* (Janich 1996: 142).

As regards the frames in the film reel that, according to the analogy, would correspond to the physicalistic time, it is not obvious how such an image could be in keeping with some puzzling questions concerning the ontology of the physical time:

- the *asymmetry* of time rendered by the metaphor of the stream and, from a scientific point of view, tested by the entropy increase (Reichenberger 2005);
- the measurement of the present in quantum mechanics (Dobbs 1969; Craig 2000).

Nevertheless, the "*laterna magica*" simile sheds light on other important questions regarding the different grammars of time. Is the future predetermined (*präformiert*)? Wittgenstein's answer to this point depends once more on the point of view taken to *interpret* the simile: the sliding frames of the projector or the *spectator*: "On the film strip there is a present picture and past and future pictures: but on the screen there is only the present" (Wittgenstein 1975: 83).

In some sense, one can know *now* which pictures will make their appearance on the screen in the *future*; it suffices, in order to have such knowledge, to get hold of the film strip. This, however, displays what has *already* happened so that talking about the "future," more precisely, about the slides which happen in quick *succession* has no sense, since such a succession *cannot be otherwise*. On the contrary, the future lying ahead of a human action cannot be *foretold* (predetermined) because, in carrying out an action, one is placed *in* time and *in* space. From the spectator's point of view, the pictures on the screen would be, in some sense, *outside* time and space. The image of the sliding frames, analogous to that of "flowing time," is rejected by Wittgenstein if by time what ought to be meant is the *possibility of change*, that is, of *being otherwise* than is expected:

> The feeling we have is that the present disappears into the past without our being able to prevent it. And here we are obviously using the picture of a film strip remorselessly moving past us, that we are unable to stop. But it is of course just as clear that the picture is misapplied: that we cannot say 'Time flows' if by time we mean the possibility of change. What we are looking at here is really the possibility of motion: and so the logical form of motion (Wittgenstein 1975: 83).

In "memory-time," the past appears as *recollection* and the future as *expectation*. If, in point of fact, memory were a *picture* representing past events, then it would *fade* as every physical picture fades and, moreover, it could be exercised as a "source of knowledge," as the *verification* of our propositions (Dobbs 1951). In spite of that, memory is an *image* and not a *picture* and, for this reason, it does not obey the rules that hold for the names of physical objects.

Grammatical misconceptions occur then when we are *tyrannized* by a metaphor and are unable to *move outside of* it.

And among the metaphorical traps which change our subject, time, there are improper language uses that lump together the *measurement* of time and the *nature of time*, as if one could say: "The height of Mont Blanc depends on how one climbs it" (Wittgenstein 1953: 225). What time means is learnt, *among other things*, by learning what it is to determine/measure time.

To make a distinction between the "measure" and "nature" of time does not mean to exclude the *Sprachspiele* in which the two conceptual terms work perfectly together. After all, in examining how we measure time or how we can get a definition of it, we merely *talk* about time. Time, in this respect, is not *in itself and for itself* since, as attested by Augustine himself, human actions, including measurement, occur in the *medium* of language: "*metimur, ut possimus dicere*" (Augustine Conf. XI 21, 27; Perissinotto 1997). Not to forget that it *takes time* to utter even a single syllable and that a key of access to time, as Augustine pointedly remarks, is given from the connection *we establish* between short and long syllables and from the *ability* to memorize the *rhythmic* succession of them. Time, in some sense, shares the same grammar as music:

> Dicturus sum canticum, quod novi: antequam incipiam, in totum exspectatio mea tenditur, cum autem coepero, quantum ex illa in praeteritum decerpsero, tenditur et memoria mea, atque distenditur vita huius actionis meae in memoriam propter quod dixi et in exspectationem propter quod dicturus sum (Augustine Conf. XI 28-38; Augustine *De Musica* 6, 8, 21; Teske 2001: 154–155).

Taking on a "liberal" attitude toward time, it is not surprising if it can be looked upon, in one respect, as a *contingent fact* regarding the world and, in another, as the most essential source of our world experience, as if we could have at our disposal two different conceptual frameworks: the first concerning a *tautological* and *formal criteria of truth*, the second regarding a *logic of content*. The Austrian philosopher nevertheless, makes no secret of preferring a characterization of time that rejects a truth-functional interpretation.

Since time and truth-functions have such different flavors and since they manifest their nature solely and completely in grammar, grammar has to explain their difference in flavor.

One tastes like content, the other like form of representation.

They taste as different as a map and a line crossing out the map (Wittgenstein 2005: 92).

As for the notion of "game" in the *Philosophical Investigations* (Wittgenstein 1953: §656), it is impossible to describe something like a *common denominator* ("essence") shared by every sentence involving time. As pointed out by Wittgenstein in his *Philosophical Grammar*, there are very different senses or uses of tensed sentence. Take as case in point the following sentences (Wittgenstein 1974: 217–218; Baker 2003: 487):

> The weather is marvelous outside.The inn flows into the Danube.Some time ago…I hope he will come.The earth was once a ball of gas.

Anyway, Wittgenstein seems to *shade off* in the more mature phase of his philosophical itinerary the opposition between grammar and content, in other words, between *syntax* and *semantics*, since he judges that empirical propositions, with time and use, can become *fossilized* assuming the form of grammatical (logical) propositions and vice versa.

It might be imagined that some propositions, of the form of empirical propositions, were hardened and functioned as channels for such empirical propositions as were not hardened but fluid; and that this relation altered with time, in that fluid propositions hardened, and hard ones became fluid (Wittgenstein 1969: §96).

The quotation above can serve as a warning to *not stiffen* concepts or conceptual relations: applied to the notion of time, this means that even the difference "physical/public-time" and "memory-time" can be *relativized* depending on contexts: "But I distinguish between the movement of the waters on the river-bed and the shift of the bed itself; though there is not a sharp division of the one from the other" (Wittgenstein 1969: §97).

The analysis of those primitive *language forms*, included the tensed uses of language, read as the pragmatic and existential background on which more complicated language systems are built.

By assuming such a philosophical attitude, Wittgenstein attempts, on the one hand, a *critical language therapy* and, on the other, the achieving of a *complete clarity* and *overall view* about the world.

When the philosopher tries to cut off the link between the primitive and the second-order language forms, he runs the risk of *generalizing* the latter, *freezing* and *objectifying* their "surface grammar" so that, for example, time becomes a *stream* or an *orientated arrow*. In so doing, the philosopher falls victim to the *Platonic obsession* of finding a *definition* for every concept or philosophical question as though it were the very lack of a definition that was responsible for our abused and mistaken linguistic uses (Kaspar and Schmidt 1992: 577). Instead of searching for a definition of time, that is, for the "ethereal essence" of it, one ought to "travel down all the lines" (Baker 2003: 487) that everyday language has drawn up for the use of this word. This travel is *disorderly*, for the ordinary use of words leaves "huge gaps" that the philosopher tries, getting it wrong, to "fill up" by asking for a *cause*, a *reason* or just a *definition*.

Consider as an example the question "What is time?" as Saint Augustine and others have asked it. At first sight what this question asks for is a definition, but then immediately the question arises: "What should we gain by a definition, as it can only lead to other undefined terms?" And why should one be puzzled just by the lack of a definition of time, and not by the lack of a definition of "chair"? Why should not we be puzzled in all cases where we haven't got a definition? Now a definition often clears up the *grammar* of a word. And in fact it is the grammar of the word time which puzzles us. We are only expressing this puzzlement by asking a slightly misleading question, the question: "What is…?" This question is an utterance of unclarity, of mental discomfort, and it is comparable with the question "Why?" as children so often ask it. This too is an expression of a mental discomfort, and does not necessarily ask for either a cause or a reason (Hertz 1899: 7). Now the puzzlement about the grammar of the word "time" arises from what one might call apparent contradictions in that grammar (Wittgenstein 1958: 26).

Our conceptual confusions arise from *grammatical contradictions* rather than from a *lack of definitions*: from blending, for example, the "stream" or "length" metaphor of time with that spotting time as "clock time." Searching for a *reason* or for a *cause* of the use of a word is not the right way to understand its sense that only a living context can give it. The "trivial" analogy between time marked as *measurement* and the same seen as, so to say, living present exerts on us a "fascination" which prevents us from distinguishing the two different grammars underlying it.

The different *paradigms* or *rules* that govern the use of tensed sentences depend ultimately on the "conceptual variety" by which the term "measurement" is characterized and it is not obvious that *length* and *clock* are the sole analogies at one's disposal.

If we look at a river in which numbered logs are floating, we can describe events on land with reference to these, e.g., "*When* the 105th log passed, I ate dinner." Suppose the log makes a bang on passing me. We can say these bangs are separated by equal, or unequal, intervals. We could also say one set of bangs was twice as fast as another set. But the equality or inequality of intervals so measured is entirely different from that measured by a clock. The phrase "length of interval" has its sense in virtue of the way we determine it, and differs according to the method of measurement. Here the criteria for equality of intervals between passing logs and for equality of intervals measured by a clock are different. We cannot say that two bangs 2 s apart differ only in degree from those an hour apart, for we have no feeling of rhythm if the interval is an hour long. And to say that one rhythm of bangs is faster than another is different from saying that the interval between these two bangs passed much more slowly than the interval between another pair (Wittgenstein 1979: 13).

In order to neutralize the philosophical problems that arise from a harmful use of words, Wittgenstein often resorts to *fictitious* and *built up* language games on the basis of which we are able to discover the deepest grammar underlying our everyday sentences. In so doing, he proves that our time sentences do not necessarily need to be anchored to length or stream metaphors that have the conceptual disadvantage of *objectifying* time and *hypostatizing* the present. One practical method of finding out such fictitious language games, real *Denkexperimente*, is to imagine how we could teach tensed sentences to children, trying to put them in *pragmatic contexts* (to tell a story, to eat, to play, to get up and so on) (Wittgenstein 1958: 81). Such living images (*Bilder aus dem Leben*) permit us to build temporal relations (*earlier*, *now*, *later*) without necessarily recurring to an objectified figure of time.

The primitive language games built up by the Austrian philosopher avoid, by way of example, employing a word such as "now" as the *name* of a moment in the stream of time.

Since 1929, when he clearly rejects phenomenological language as *absurd*, favoring the physicalistic one as the sole legitimate, the Austrian philosopher, even though he acknowledges different time notations, is urged to see them as *different dialects* of one basic physicalistic language. In *Philosophical Remarks*, phenomenological *Denkexperimente* come to a "parasitic variety" (Hintikka 1996: 246) within a more general physicalistic framework. For this reason, phenomenological language is unfit to assume the role of a primary or basic framework by virtue of which, for example, to express *directly* the living time. A basic phenomenological language, actually, would be "that inarticulate sound with which many writers would like to begin philosophy" (Wittgenstein 1975: 98).

In short, even if allowed, phenomenological languages of any sort cannot be taken as independent, standing on their own feet, or frameworks.

Such a theoretical result, though, is not completely satisfying insofar as the legitimate distinction between a phenomenological and a physical discourse falls down to the contrast between a *perspectival* and a *public* framework. As observed by Hintikka (1996: 270–274), the real issue pertaining to the "dialectic" between different language frameworks would be that of the *integration* of all *local perspectival* systems (space, time, and so on) into a single public system. This was the problem Russell was confronted with in his *theory of acquaintance*.

This way of looking at the problem would not create a problem to the *coexistence* in the everyday language of phenomenological and physicalistic frames of reference. How can one *compare* local perspectival and physicalistic languages? Even if our ordinary language is mostly geared to a public system of identification, it would be a mistake to consider locally perspectival frameworks as dependent or parasitic on the former. Concepts tailored to perspectival identification as "this," "that," and even "now" do work perfectly well as those requiring the public framework. *Definite descriptions* argue for contexts' dependent values and even *quantifiers*, and the logic supporting them, can be brought back to a perspectival dimension. All this, to conclude that Wittgenstein's identification of the phenomenological frame with the perspectival one is wrought with problems. This negatively affects the use of the term "memory-time" as if the only viable determination of time-references were that founded on *momentary epistemic* states of memory, that is, on merely *internal criteria*. But what if we mean by "memory-time" *personal episodic memory*, that is, something which is tied to a spatiotemporal perspective, to my "world line" in space-time? With respect to this framework, the idea that in memory-time there are only "earlier" and "later" relations or the view that it would be impossible to measure remembered time-spans would be weird. After all, to measure or compare remembered time-spans is not very different from comparing two perceived spans in the visual space.

If so, we can argue that "memory-time" does not necessarily involve a special relation to memory. It sounds as if the Austrian philosopher had burdened the concept of time with a *psychologistic emphasis*. The "depersonalization" of time-reference, no more related to memory as an *intrinsically* private source of identification, would have another important consequence insofar as it would allow us to bring in even questions regarding the explanation of space as requested by the *theory of relativity*. In fact, the integration of different (spatial) perspectival frameworks into a single public one takes into consideration also the integration of different local perspectival times into an *absolute* one, a question that was of particular interest in Einstein's special theory of relativity.

Wittgenstein and Einstein, as regards the issue of time relativity, certainly present some similarities—for example, the connection between measurement method and nature of an object—but these break down insofar as the first does not go *beyond* the grammar of everyday language, while the second *substantially* modifies our *manifest image* of the world and he would have not subscribed to the view that "there is no more difficulty about time than there is about this chair" (Wittgenstein 1979: 119).

Besides, a Sellarsian approach to the question of time for which time is a kind of a *theoretical entity* "postulated" already within the framework of everyday observational discourse and "refined" by science so that events in (physical) time can be seen "as metrical abstractions grounded in the reality of changing substances" (Sellars 1975), would have been more promising than that adopted by Wittgenstein in accordance with the fact that the former allows for an "amalgamation" of social time, including also the living present, with the time of physics (Nyíri 2006).

Even though Wittgenstein's remarks on time are scattered throughout his work, they do not make up a systematic corpus. Nevertheless, they are far-reaching since they invite us to reflect on this subject without philosophical bias. Following Wittgenstein's fashion of thinking, we can formulate some questions about time and try to answer them by quoting Wittgenstein himself.

(a) Is time a *real* and *objective* entity?

"*After all, time is not a temporal space, but an ordering*" (Wittgenstein 2005: 363).

Does the indexical "now" in the sentence "The moon is now rising" convey that the moon's rising has the property of presentness (Smith 1993) or does the indexical "now" simply refer to the date or time of the moon's rising—or, at least, to the fact the it stands in some temporal relation—(Mellor 1981; Oaklander 1990)?

(b) Is time constituted by particular *relations* of "precedence" or "succession" among temporal elements (*moments* or *events*) or is it founded on peculiar *properties* called "past," "present," "future"?

"One can talk about present, past and future events in the physical world, but not about present, past and future mental images" (Oaklander 1990). Only as related to physical events, time can have a *tense dependent* syntax. Considered as a mental image (of memory), time is confined to the present and has no *alethic value* since "to speak of memory as the source of our cognition, as the verification of our propositions, has to lead to nonsense" (Oaklander 1990).

(c) Is it possible to talk about time in *modal terms*, that is, employing possible worlds semantics?

> Can I say that drama has its own time that is not a section of historical time? That is to say, within a drama I can speak about before and after, but the question whether the events took place, say, before or after Caesar's death makes *no sense* (Wittgenstein 1980: 13 e).

One way, and perhaps not the "correct" one, of tying such different times could be that of turning to a notion of time built on the indexical framework ("earlier," "later") of "memory-time", intended as the source of living time (Orilia 2012). Moreover, a modal understanding of time has also ethical repercussions insofar as it insinuates "*futura contingentia*" (*à la* Duns Scotus) and therefore the plausibility of *liberum arbitrium*:

> [...]. Neither can we say—and here is another source of misunderstanding—"It's now the case that this event will occur in an hour" or "It's the case at 5 o'clock that I'll take a walk at 7 o'clock" (Wittgenstein 2005: 364).

(d) Do "ordinary objects" such as chairs, computers, trees, and the like *persist* in time in the sense of *tridimensionalism* (Mellor 1998: chap. 8) or *quadridimensionalism* (Van Orman Quine 1960: §36)?

For instance, is the PC that I am *now* using an object that, in some way, is, *at the same time*, "past" (for it already existed) and "future" (for it will *probably* exist tomorrow)? And what can we add about events that are "repeatable" like the revolutions of the clock hands?

Drawing on Wittgenstein, we could say that the adoption of a "clock grammar" might be *consistent* with the idea of the repeatability of events, in contrast to McTaggart's principle of the *irreality* of time (McTaggart 1908; Orilia 2012), while it could prove to be *inconsistent* with a physicalistic or a *verificationist* approach to time:

> How language distances itself from a description of verification. How abstract it gets! We have to rediscover that we measure time with a clock. And in the process we do not even notice that we have made a grammatical discovery (Wittgenstein 2005: 208e).How a proposition is verified, that's what it says. […]. Verification isn't merely an indication of the truth; it determines the sense of the proposition. (Einstein: How a quantity is measured, that's what it is). (Wittgenstein 2005: 208e).

Quoting Hannah Arendt's brilliant interpretation of the *style* or *color* of Wittgenstein's trains of thoughts, this paper aims, dear reader, at encouraging you to *stop and think* about time. After all, in so doing we transform the activity of thinking into a *breathless*, *rhythmical,* and therefore, strongly *temporal* movement of inquiring (Arendt 1977: 231 n.133).

References

Arendt, H. (1977). *The Life of the Mind*. San Diego: Harcourt Brace Jovanovich.

Augustine. *Confessionum*. http://www.augustinus.it/latino/confessioni/index2.htm.

Augustine. *De Musica*. http://www.augustinus.it/latino/musica/index2.htm.

Baker, G. (Ed.). (2003). *The Voices of Wittgenstein*. Routledge: The Vienna Circle.

Craig, W. L. (2000). *The Tenseless Theory of Time. A Critical Examination*. Dordrecht: Kluwer.

Dobbs, H. A. C. (1951). The relation between the time of psychology and the time of physics, Part I. *British Journal for Philosophy of Science, 2*(6), 122–141.

Dobbs, H. A. C. (1969). The 'present' in physics. *The British Journal for the Philosophy of Science, 19*(4), 317–324.

Hintikka, J. (1996). *Ludwig Wittgenstein. Half-Truths and One-and-a-Half-Truths*. Dordrecht: Springer.

Hertz, H. (1899). *The principles of mechanics*. London: MacMillan.

Kaspar, F. R., & Schmidt, A. (1992). Wittgenstein über Zeit. *Zeitschrift für philosophische Forschung, 46*(4), 569–583.

Malcolm, N. (2001). *Ludwig Wittgenstein. A Memoir*. Oxford: Clarendon Press.

McTaggart, J. M. E. (1908). The Unreality of Time. *Mind, 17*, 457–474.

Mellor, D. H. (1981). *Real Time*. Cambridge: Cambridge University Press.

Mellor, D. H. (1998). *Real Time II*. London: Routledge.

Moore, G. E. (1993). Wittgenstein's Lectures in 1930–33. In: C. J. Klagge, & A. Nordmann (Eds.), *Ludwig Wittgenstein. Philosophical Occasions* (pp. 46–114). Cambridge: Hackett.

Nyíri, K. (2006). Time and communication. In F. Stadler & M. Stölzner (Eds.), *Time and History* (pp. 301–316). Heusenstamm: Ontos Verlag.

Oaklander, L. N. (1990). The new tenseless theory of time: a reply to Smith. *Philosophical Studies, 58*, 287–292.

Orilia, F. (2012). *Filosofia del tempo. Il dibattito contemporaneo* (pp. 99–113). Roma Carocci.

Perissinotto, L. (1997). *Wittgenstein. Una guida*. Milano: Feltrinelli

Peter, J. (1996). Die Konstitution der Zeit durch Handeln und Reden. *Kodikas/Code: Ars Semeiotica, 19*, 133–147.

Reichenberger, A. A. (2005). Was ist Zeit?. Wittgensteins Kritik an Augustinus kritische betrachtet. In: F. Stadler, & M. Stöltzner (Eds.), *Papers of the 28th IWS (International Wittgenstein Symposia)*, (pp. 253–255). Kirchberg am Wechsel: Austrian Ludwig Wittgenstein Society.

Russell, B. (2009). *Our Knowledge of the External World*. London: Routledge.

Sellars, W. (1975). Autobiographical Reflections. In: H.N. Castañeda (Ed.), *Knowledge and Reality: Critical Studies in Honor of Wilfrid Sellars*, in Action (pp. 277–293). Indianapolis: Bobbs-Merrill.

Smith, Q. (1993). *Language and time*. New York: Oxford University Press.

Teske, R. (2001). Augustine's philosophy of memory. In Eleonore Stump & Norman Kretzmann (Eds.), *The Cambridge Companion to Augustine* (pp. 148–158). Cambridge: Cambridge University Press.

Van Orman Quine, W. (1960). *Word and Object*. Cambridge: MIT Press.

Wittgenstein, L. (1953). *Philosophical Investigations*. Oxford: Blackwell.

Wittgenstein, L. (1958). *The Blue Book*. Oxford: Blackwell.

Wittgenstein, L. (1961). *Tractatus Logico-Philosophicus*. London: Routledge & Kegan Paul.

Wittgenstein, L. (1969). *On Certainty*. Oxford: Blackwell.

Wittgenstein, L. (1974). *Philosophical Grammar*. Oxford: Blackwell.

Wittgenstein, L. (1975). *Philosophical Remarks*. Oxford: Basil Blackwell.

Wittgenstein, L. (1979). *Lectures 1932–1933*. Oxford: Blackwell.

Wittgenstein, L. (1980). *Culture and value*. Oxford: Blackwell.

Wittgenstein, L. (2005). *The Big Typescript: TS 213*. Oxford: Wiley/Blackwell.

Sign(s) of the Time: Time and Understanding in Heidegger's Phenomenological–Ontological Hermeneutics

Simona Venezia

Abstract The paper discusses the relationship between time and understanding in Heidegger's phenomenological-ontological hermeneutics. Even thanks to an innovative concept of understanding as an open and projecting dimension, Heidegger can reach the qualitative, dynamic and differential concept of time, which is the basis of the Daseinsanalyse in Sein und Zeit. Only if comprehension is meant as a primary phenomenon can time be thought as an ecstatic disclosedness, i.e. an original, ontologically inderivative, unprogrammable non-functionalistic and essential temporality, which always involves and concerns us.

In the broad framework of Heidegger's question about time, this paper discusses only one specific perspective, i.e., the relationship between time and understanding. We start from the assumption that a new relationship between time and understanding is one of the most revolutionary discoveries of *Sein und Zeit* (Heidegger 2001), the masterpiece of 1927 with which Heidegger established himself on the international academic scene by reviving the question about time in philosophical research.

The project of *Sein und Zeit* has a dual purpose: on the one hand to raise the question about being outside an ontic theory, and on the other to identify time with being. The title *Being and time* could be understood even as *Being is time*: only with the identification of being with time is it really possible to remove from being every ontic characterization in order to think of it in a fully ontological horizon. This claim is based on an innovative concept of time: time is no longer intended as a fixed and stable permanence, as a substantialistic and incontrovertible foundation (as in the conceptual-philosophical thought and in the ordinary way of thinking), but as an ecstatic interaction between past, present, and future, an openness to the irrevocable human finitude that allows the subjectivity to grasp the ultimate truth of its existence.

S. Venezia (✉)
Philosophy Section, Department of Humanities, University of Naples Federico II, via Porta di Massa 1, 80133 Naples, Italy
e-mail: simona.venezia@unina.it

F. Santoianni (ed.), *The Concept of Time in Early Twentieth-Century Philosophy*, Studies in Applied Philosophy, Epistemology and Rational Ethics 24,
DOI 10.1007/978-3-319-24895-0_17

Heidegger's meditation about time starts from the harsh critique against the traditional concept belonging to western metaphysics, that from the beginning has reduced time into a kinetic-numerical-quantitative concept by following the famous definition from Aristotle's *Physics* of time as «the number of movement in respect of *before* and *after*» (see Aristotle 1996, IV 129b). The German philosopher intends to replace the quantitative concept of time with a qualitative one: moving from this need he finds the ontic-ontological as well as phenomenological–hermeneutical *Zeitlichkeit*. The *Zeitlichkeit* does not indicate a general, comprehensive concept of time, but a very precise one, i.e., a concept bound to the *Facticity* [*Faktizität*][1] of existence. *Faktisch* indicates the radical temporal finitude of the human being, that expresses itself in the inescapability of our Being-toward-death [*Sein zum Tode*].[2] Time is not *what* we live, but *how* we live. In this new perspective, Bergson and his reflection about time as *duration* [*durée*] (see Bergson 1889: 75–139) played a very important role: despite the fact that Heidegger strongly criticized the concept of duration because of its belonging to metaphysics, it is indisputable that Bergson's analyses put in question precisely the quantitative matrix of the time as mere extension proposed by the natural sciences.

Heidegger's purpose is therefore to think of time as a qualitative and not quantitative dimension (see Heidegger 2011), dynamic and not static, differential and not uniform, and no longer as a stable and permanent present. In this way, it is possible to overcome a concept of time as a secure foundation in order to think of an original eventuality, an ecstatic openness (never predictable or able to be passed through) of past, present, and future. These three «*ecstases*» [*Ekstasen*][3] can no longer be measured according to a criterion of succession. The succession assures us by giving stability and providing certainty: if our existence flows linearly, we can have no doubt about its consistency and legitimacy. The metaphysical concept of time satisfies our need for reassurance, but does not reveal the truth of our existence. The opposition to this need shows the necessity to think of being and time together: just as time, being is an original, unrepresentable, and uncategorizable eventuality.

With his reflection about time, Heidegger intends to overcome the theoretical categories that alienate the philosophical analyses by the *pathos* of existence. The confrontation with the history of western metaphysics is central but not sufficient because of the lack of 'pathicity' of its horizon. For this reason, the German philosopher turned to the Christian experience by analyzing the *Urchristentum* of

[1] «"*Facticity*" is the designation we will use for the character of the being of "our" "own" *Dasein*. [...] "*factical*" means something which is of itself articulated with respect to, on the basis of, and with a view to such a *factical* character of being and "is" in this manner» (Heidegger 1999a: 5).

[2] «In Being-towards-death, Dasein comports itself *towards itself* as a distinctive potentiality-for-Being» (Heidegger 2001: 296).

[3] «*Temporality is the primordial 'outside-of-itself' in and for itself.* We therefore call the phenomena of the future, the character of having been, and the Present, the "*ecstases*" of temporality. Temporality is not, prior to this, an entity which first emerges from itself; its essence is a process of temporalizing in the unity of the ecstases» (Heidegger 2001: 377).

the primitive communities (see Heidegger 2004b). Even if Heidegger finds in the Augustinian concept of time as "extension in the mind (soul)" [*distentio animi*] (see Augustine 2012, XI, 26), the possibility to overtake a traditional concept of time as a measurable objectivity thanks a new dimension of interior spatiality, he is much more inspired by the eschatological kairology of primitive Christianity. Only kairology can overcome chronology. Only the kairological time, that waits for salvation, can overcome the chronological time of programmable certainty. For the primitive Christians, the event of salvation cannot be predicted, planned, or calculated: it will happen in a *kairós*, a decisive moment that men cannot schedule, but only wait for. The *phenomenology of religious life* shows that this expectation does not provide an object because it is an event of openness. This type of event cannot be understood by metaphysical theoreticism: only an innovative philosophical approach can reach it. This new orientation is offered by the *Daseinsanalyse* of *Sein und Zeit*, in which Heidegger proposes a phenomenological–ontological hermeneutics, following a program that began with the so called *Natorp-Bericht* (Heidegger 2005) and the lecture course from the summer semester of 1923 (Heidegger 1999a) and that aimed at overcoming every theoreticism, logicism and vitalism, psychologism and historicism belonging to contemporary philosophy. The hermeneutical matrix of the *Daseinsanalyse* allows Heidegger to propose an antisubjectivistic ontology that is at the same time antitheoretical and antipsychologistic: the *Facticity* of existence indicates the concreteness of life in its radical and original temporality.

This ontology applies a phenomenological approach because the *Seiende* is analyzed in its manifestation in a context (see Heidegger 1979). At the same time, it applies also a hermeneutical approach because the context proves to be an originally meaningful context. This hermeneutical matrix is based on a new, revolutionary concept of understanding [*Verstehen*]: «*Understanding is the existential Being of Dasein's own potentiality-for-Being; and it is so in such a way that this Being discloses in itself what its Being is capable of*» (Heidegger 2001: 184). If we think the *Verstehen* as an *Existentiale*, we refuse a concept of understanding as an intellectual operation, a cultural performance, and above all as an external intervention which the subject produces. Only in this way does understanding become an essential ontological openness which connects subjectivity to the world. Heidegger is not interested in the subject [*Subjekt*] of traditional western metaphysics (see Heidegger 1977b and Heidegger 1998), which stands in front of the world like in front of a foreign object, which can be judged and known in its incontrovertible essence. The relationship between subject and object is totally eliminated by the *Being-there* of the *Dasein*, our *Being-in-the-world* [*In-der-Welt-sein*] that is *already* open to the world. This openness is never neutral, because it affects our primary relationship with the world: this relationship is never a pure perception, but an interpreting comprehension [*auslegendes Verstehen*]. We live in a world that is already meaningful, not a simple collector of people, things, and occurrences. Heidegger proposes a new concept of the *Welt*: the world is not a place containing something or someone, but an open context capable of activating meaningful connections, a «complex of relations» (Richardson 2003: 63). Only an ontological hermeneutics can understand what the world really is, because this philosophical perspective does not look for eternal essences and substantialistic foundations, but it aims at outlining the net of sense connections that makes up our being in the world. Only ontologically it is possible to grasp the *Verstehen*: the world cannot be understood through knowledge categories, because we are already placed in this world we want to interpret. The ontological issue of this phenomenological hermeneutics is granted by the structure of the preliminary understanding, the *pre-comprehension* [*Vorverständigung*][4] (Heidegger 2001: 15) meant as the negation of every transcendental assumption: when we understand, we are actually *already* surrounded by a world capable of activating a *pre-understanding*, which, even if not yet schematized and categorized, affects our understanding. We comprehend only because we are open to the world we live in and we want to interpret: «the phenomenology of Dasein is a *hermeneutic* in the primordial signification of this word, where it designates this business of interpreting» (Heidegger 2001: 62). The phenomenon of interpretation is not derivative, subsequent as compared to *Dasein*'s openness to the world, but is contextual and above all constitutive of this openness.

If we want to understand this theoretical effort, we have to consider the close link (which Heidegger will always take into account, even after the so called *Kehre*) between the disclosedness [*Erschlossenheit*][5] and the "to-be-able-to-be," i.e., the "Being-possibile" [*Seinkönnen*]. As we said, *Dasein* is a dimension open to the world context: this disclosedness is based on the fact that this new type of subjectivity is no longer considered as a reality [*Wirklichkeit*], a fixed and unwavering real subject, but a potentiality [*Möglichkeit*],[6] a to-be-able-to-be that constitutes itself while it constitutes sense in the world context: «In understanding, as an *Existentiale*, that which we have such competence over is not a "what", but Being as existing. The kind of Being which Dasein has, as potentiality-for-Being, lies existentially in understanding. Dasein is not something present-at-hand which possesses its competence for something by way of an extra; it is primarily Being-possible. Dasein is in every case what it can be, and in the way in which it is its possibility» (Heidegger 2001: 183).

Starting from this new concept of understanding as potentiality (belonging to the context of the meanings of the world), Heidegger therefore proposes an innovative definition of subjectivity, the *Dasein* outlined above (that is actually no longer a subjectivity), which overcomes the relationship between subject and object: «As understanding, Dasein projects its Being upon possibilities» (Heidegger 2001: 188). In this way, we can finally focus attention on the link between *Dasein* and *Verstehen*, i.e., the *project* [*Entwurf*]: «The projecting of the understanding has its

[4]In the English edition the term *Vorverständigung* has been translated in a generic and imprecise way as «understanding» (Heidegger 2001: 31).

[5] «*Dasein is its disclosedness* [*Daseins is seine Erschlossenheit*]» (Heidegger 2001: 171).

[6]One of the most important thesis present in *Sein und Zeit* claims that: «Higher than actuality stands possibility [Höher als die Wirklichkeit steht die *Möglichkeit*]» (Heidegger 2001: 63).

own possibility-that of developing itself [*sich auszubilden*]. This development of the understanding we call "interpretation"» (Heidegger 2001: 188). Heidegger's ontological hermeneutics is therefore a *Hermeneutics of project* (see Venezia 2014a), which considers the subjectivity as an understanding and as an interpreting Being-possible. This potentiality does not refer to any crystallized reality, because it does not decide any truth about the surrounding world, but it is open to the sense of this world. From this point of view, it is essential to underline the importance given to the dimension of the *future*. If we really intend to think of an unrepresentable and uncategorizable dimension of time, we have to focus on the priority that Heidegger assigns to the future in all his works (see Venezia 2014b): this priority appears to be fundamental already in the phenomenology of religious life from the early Twenties in which eschatology does not outline the expectation of the end, but means to-be-ready for a salvation coming without warning and without scheduling by living deeply the extreme finitude of our facticity. In the *Daseinsanalyse,* the primacy of the future is necessary to think of the *Sein zum Tode*: we are used to thinking of death starting from time, but now we have to think of time starting from death and from our extreme finitude. In the same manner, we must no longer consider the future starting from time, but time starting from future. Even in the so called 'second Heidegger' the primacy of the future is central: the most proactive part of the entire project of the *Contributions to Philosophy* is based on the *fugue* [*Fuge*] of "The Ones to Come" [*Zukünftigen*] (Heidegger 1999b: 277–281), who qualify their time in the expectation of the *last God*. This "God" does not indicate a religious utopia, but means a new relationship with time: the *Zukünftigen* qualify their existences by living deeply the expectation without a stable, transcendent foundation. Future is not a supernatural dimension, different from the life we live in, but the limit that configures our lives. For this reason, Heidegger's reflection about time and future reaches its apex in the famous sentence: «But origin always comes to meet us from the future» [*Herkunft aber bleibt stets Zukunft*] (Heidegger 1971: 10). Within the identification of origin [*Herkunft*] and future [*Zukunft*], it is definitely possible to overcome a concept of time as chronological quantity in order to propose a concept of time radically thought as an incalculable and original event.

Even if not yet capable of focusing on the central role of the *event* for the question of time, the phenomenological–ontological hermeneutics of *Sein und Zeit* aims at proposing a wide-perspective reflection and not only one that referring to ontological themes. In the paragraph 17, entitled *Reference and signs* [*Verweisung und Zeichen*], Heidegger deals with a very important question for international linguistics after Peirce and Saussure, i.e., the question of *sign*. The German philosopher makes an essential modification: while for Peirce sign «or *representamen*» is «something which stands to somebody for something in some respect or capacity» (Peirce 1931–1935: CP 2.228), for Saussure the linguistic sign is «a two-sided psychological entity» that «unite[s], not a thing and a name, but a concept and a sound-image» (Saussure 1959: 66), according to Heidegger sign is above all an *equipment* [*Zeug*] (Heidegger 2001: 107). This definition is almost pragmatic, but Heideggerian pragmatism is different from the traditional one, because it is based on one of the most important and revolutionary concepts of *Sein und Zeit*, i.e., the «*readiness-to-hand*» [*Zuhandenheit*].[7] With this concept, it is possible to overcome the metaphysical tradition that considers the «constant presence-at-hand» [*Vorhandenheit*] (Heidegger 2001: 129) the way the subject is related to the world. In the *Zuhandenheit,* Heidegger finally finds a philosophical foundation for the concept of *sign*, that neither Peirce nor Saussure could find (also because they did not want to). They focus their attention only on the *Vorhandenes*, without realizing that a simple presence cannot activate a signing force. Only with the *Zuhandenes,* is it possible to understand that a sign is what refers to something other in a context of sense and what links the different subjectivities in their being in the world. The great Heideggerian breakthrough consists in the fact that what is meaningful [*das Bedeutsame*] becomes what is primary [*das Primäre*]. This primary cannot be perceived through a mental acquisition, but through an interpretation. In the context of the world, things surround us as already meaningful: Husserl believed that this meaningfulness had its origin in the subjectivity, while Heidegger believes that this meaningfulness has its origin neither in subjectivity nor the world, but in the context of subjectivity and world, in their interaction, in their contextual and constitutive openness. The question of *significance* [*Bedeutsamkeit*] becomes therefore absolutely central: in *Ontology. Hermeneutics of Facticity*, Heidegger already claims that significance is «the character of the world's being-encountered» (Heidegger 1999a: 71); this *Begegnischarakter* overtakes the *Subjekt-Objekt* relationship because the encounter itself is understood as a primordial, inderivative phenomenon. In *Sein und Zeit,* significance is also defined as the structure of the world [*Struktur der Welt*] (Heidegger 2001: 120): not a tool, an accessory, a subsequent attribute, but the constitution of *Dasein*'s identity. The *Bedeutsamkeit* allows to recognize in the *verstehende Auslegung* the fundamental relation between *Dasein* and the world: «In the act of *understanding* [...] the relations [...] must have been previously disclosed; the act of understanding holds them in this disclosedness. [...] These relationships are bound up with one another as a primordial totality; they are what they are as this signifying [Be-*deuten*] in which Dasein gives itself beforehand its Being-in-the-world as something to be understood. The relational totality of this signifying we call "*significance.*" This is what makes up the structure of the world-the structure of that wherein Dasein as such already is» (Heidegger 2001: 120).

Heidegger theorizes the primariness of the *verstehende Auslegung* upon the *Aussage*, that corresponds to the primariness of the *Zuhandenheit* upon the *Vorhandenheit*; before every assertion, in fact, there is the interpretation: «But in significance itself, with which Dasein is always familiar, there lurks the ontological condition which makes it possible for Dasein, as something which understands and interprets, to disclose such things as 'significations'; upon these, in turn, is founded the Being of words and of language. The significance thus disclosed is an existential state of Dasein-of its Being-in-the-world; and as such it is the ontic condition for the possibility that a totality of involvements can be discovered» (Heidegger 2001: 120).

---

[7]«The kind of Being which equipment possesses—in which it manifests itself in its own right—we call "*readiness-to-hand*" [*Zuhandenheit*]» (Heidegger 2001: 98).

In this way, Heidegger definitively leaves Husserlian phenomenology (see Heidegger 1982). He poses in fact a decisive question for realizing the passage from phenomenology to hermeneutics, in consideration of the fact that this passage can realize itself only in ontology: «But to what extent does it become a derivative mode of interpretation [*abkünftiger Modus der Auslegung*]?» (Heidegger 2001: 200). The answer is based on the fact that «interpretation [*der ursprüngliche Vollzug der Auslegung*] is carried out primordially not in a theoretical statement» (Heidegger 2001: 200). According to a very solid tradition, thought begins from the assertion subsequent to perception: between perception and assertion a complex and partially mysterious phenomenon of comprehension would occur. After this understanding comes the interpretation: this phenomenon is understood as a filter, a decoding of something essentially unknown, that reveals itself as something knowable once it is decoded by human cognitive-intellectual skills. In the light of this, Heidegger's discovery consists in unifying two separate phenomena, i.e., understanding and interpretation, and in showing that the interpretation is not a derivative of the assertion, but that the assertion is a derivative of the interpretation.

The assertion is not therefore the primary expression of thought: it cannot exist without the interpreting life of *Dasein* in the world context. In this way Heidegger can finally overcome the theoreticism by showing that it is actually 'only' a construction, a superstructure, something artificial, and subsequent in confrontation with our primordial contact to the world.

It is thanks to this new way of thinking the comprehension that Heidegger can really think of time in an innovative way. Only if understanding does not mean the possession of a thought as we possess an object, but means openness to what is different from us, is time no longer thinkable as a measurable quantity. Time must be understood as an ecstatic disclosedness, which always involves and concerns us; above all as an original, ontologically inderivative, unprogrammable nonfunctionalistic, and essential temporality. Only in this way can the *understanding of time* be finally thought as what it really is: the *time of understanding*.

References

Aristotle. (1996). *Physics* (R. Waterfield, Trans.). Oxford: Oxford University Press.

Augustine. (2012). *Confessions*. Oxford: Oxford University Press.

Bergson, H. (1889). Essai sur le données immeédiates de la conscience. Paris: Presses Alcan. English edition: Bergson, H. (2001). *Time and Free Will: An Essay on the Immediate Data of Consciousness* (F. L. Pogson, Trans.). Mineola: Dover Publications.

Heidegger, M. (1976). *Wegmarken, Gesamtausgabe Bd 9*. Frankfurt: Klostermann. English edition: Heidegger, M. (1998). *Pathmarks* (W. McNeill, Trans.). Cambridge: Cambridge University Press.

Heidegger, M. (1977). *Sein und Zeit, Gesamtausgabe Bd. 2*. Frankfurt: Klostermann. English edition: Heidegger, M. (2001). *Being and Time* (J. Macquarrie & E. Robinson, Trans.). Oxford: Blackwell.

Heidegger, M. (1979). *Prolegomena zur Geschichte des Zeitbegriffs, Gesamtausgabe Bd. 20*. Frankfurt: Klostermann. English edition: Heidegger, M. (1985). *History of the Concept of Time*. Prolegomena (Th. Kisiel, Trans.). Bloomington: Indiana University Press.

Heidegger, M. (1985). *Unterwegs zur Sprache, Gesamtausgabe 12*. Frankfurt: Klostermann. English edition: Heidegger, M. (1971). *On the Way To Language* (P.D. Hertz, Trans.). New York: Harper and Row.

Heidegger, M. (1988). *Ontologie (Hermeneutik der Faktizität), Gesamtausgabe Bd. 63*. Frankfurt: Klostermann. English edition: Heidegger, M. (1999a). *Ontology: The Hermeneutics of Facticity* (J. van Buren, Trans.) Bloomington: Indiana University Press.

Heidegger, M. (1989a). *Die Grundprobleme der Phänomenologie, Gesamtausgabe Bd. 24*. Frankfurt: Klostermann. English edition: Heidegger, M. (1982). The basic problems of phenomenology (A. Hofstadter, Trans.). Bloomington: Indiana University Press.

Heidegger, M. (1989b). *Beiträge zur Philosophie (Vom Ereignis), Gesamtausgabe 65*. Frankfurt: Klostermann. English edition: Heidegger, M. (1999b). Contributions to Philosophy (From Enowning) (P. Emad & K. Maly. Trans.). Bloomington: Indiana University Press.

Heidegger, M. (1995). *Phänomenologie des religiösen Lebens, Gesamtausgabe Bd. 60*. Frankfurt: Klostermann. English edition: Heidegger, M. (2004b). *The Phenomenology of Religious Life* (M. Fritsch & J. A. Gossetti-Ferencei. Trans.), Bloomington: Indiana University Press.

Heidegger, M. (2004a). *Der Begriff der Zeit, Gesamtausgabe Bd. 64*. Frankfurt: Klostermann. English edition: Heidegger, M. (2011). The concept of time. *The First Draft of Being and Time* (I. Farin & A. Skinner, Trans.). London: Continuum.

Heidegger, M. (2005). *Phänomenologische Interpretationen ausgewählter Abhandlungen des Aristoteles zu Ontologie und Logik, Gesamtausgabe Bd. 62*, Frankfurt: Klostermann.

Heidegger, M. (2007). *Zur Sache des Denkens, Gesamtausgabe 14*. Frankfurt: Klostermann. English edition: Heidegger, M. (1977b). On time and being (J. Stanbaugh. Trans.). New York: Harper and Row.

Peirce, C. S. (1958). *The collected papers of Charles Sanders Peirce*. (1931–1935). In: Charles & P. Weiss (Eds.) Vol. I–VI H. and W.B. Arthur (Ed.) Vols. VII–VIII. Cambridge (Mass.): Harvard University Press.

Richardson, William J. (2003). *Heidegger: through Phenomenology to Thought*. New York: Fordham University Press.

Saussure, F. d. (1995). *Course de linguistique générale*. B. Charles & A. Sechehaye (Eds.) avec la collaboration de Albert, Riedlinger. Paris: Éditions Payot & Rivages. English edition: de Saussure, F. (1959). *Course in General Linguistics*. In: B. Charles & Albert Sechehaye (W. Baskin, Trans.). New York: McGraw-Hill.

Venezia, S. (2014a). Ermeneutica del tempo in Martin Heidegger. In F. Santoianni (Ed.), *La filosofia nello spazio del pensiero* (pp. 79–88). Roma: Carocci.

Venezia, S. (2014b). Dal fenomeno all'evento: Heidegger filosofo del futuro. In *La coscienza del futuro*. In: R. Peluso & R. V. Cavaliere (Eds.) (pp. 271–290). Napoli: La Scuola di Pitagora.

Psychoanalysis at the Test of Time: Jacques Lacan's Teaching

Marco Castagna

Abstract The question of "time" is a worthy topic of psychoanalytic inquiry, concerning at least three different levels: *technical* (relating to the *duration* of therapy), *epistemological* (relating to the *evolution of* and *between* diseases), and *existential* (relating to the analyzing subject's *consciousness* of himself). However, if subjective *well-being* is the ultimate purpose of the analytic experience, it is immediately clear that the last level also defines the previous ones. In this perspective, we can attribute an innovative role to Lacan's inquiry on time, in receiving the Freudian psychoanalytical legacy and in participating in the contemporary thought about subjective consciousness of time. Indeed, starting from the paper entitled "Logical time" (1945), Lacan completely abandons a linear notion of time: what concerns psychoanalysis is not the "real" past sequence of events in themselves, but the way these events exist "now" in memory, and the way the patient reports them. This dynamic idea of subjective temporality (indebted to Heideggerian hermeneutics as well as to Peircian semiotics) was immediately received not only by clinical therapy but also by other branches of knowledge (particularly by theory of literature, as the basis for the concept of "narrative plot"—see Brooks). Nevertheless, in his late teaching Lacan directly engages the epistemological level of time analysis, trying to offer constants to the dynamic of time consciousness. Therefore, in the last years of his *Seminar*, he introduces into his work the mathematical concept of *topological space*, and begins to delineate a close relationship between time and space from the immanent point of view of discourse theory. If "rhythm" seems to be the ultimate horizon opened by Lacan's teaching on time toward the end of his career, this perspective could mark the definitive (creative) return to Freud and anticipate some of the current most valid research in semiotics and physics. Unfortunately, the death of the French psychoanalyst resulted in this perspective remaining only barely sketched out.

M. Castagna (✉)
Philosophy Section, Department of Humanities, University of Naples Federico II, via Porta di Massa 1, 80133 Naples, Italy
e-mail: marco.castagna@gmail.com

F. Santoianni (ed.), *The Concept of Time in Early Twentieth-Century Philosophy*, Studies in Applied Philosophy, Epistemology and Rational Ethics 24,
DOI 10.1007/978-3-319-24895-0_18

The question of "time" is a worthy topic of psychoanalytic inquiry, concerning at least three different levels: *technical* (relating to the *duration* of therapy), *epistemological* (relating to the *evolution of* and *between* diseases), and *existential* (relating to the analyzing subject's *consciousness* of himself). However, if subjective *well-being* is the ultimate purpose of the analytic experience, it is immediately clear that the last level also defines the previous ones. In this perspective, we can attribute an innovative role to Lacan's inquiry on time, in receiving the Freudian psychoanalytical legacy and in participating in the contemporary thought about subjective consciousness of time. Indeed, starting from the paper entitled "Logical time" (1945), Lacan completely abandons a linear notion of time: what concerns psychoanalysis is not the "real" past sequence of events in themselves, but the way these events exist "now" in memory, and the way the patient reports them. This dynamic idea of subjective temporality (indebted to Heideggerian hermeneutics as well as to Peircian semiotics) was immediately received not only by clinical therapy but also by other branches of knowledge (particularly by theory of literature, as the basis for the concept of "narrative plot"—see Brooks). Nevertheless, in his late teaching Lacan directly engages the epistemological level of time analysis, trying to offer constants to the dynamic of time consciousness. Therefore, in the last years of his *Seminar*, he introduces into his work the mathematical concept of *topological space*, and begins to delineate a close relationship between time and space from the immanent point of view of discourse theory. If "rhythm" seems to be the ultimate horizon opened by Lacan's teaching on time toward the end of his career, this perspective could mark the definitive (creative) return to Freud and anticipate some of the current most valid research in semiotics and physics. Unfortunately, the death of the French psychoanalyst resulted in this perspective remaining only barely sketched out.

The question of "time" is a worthy topic of psychoanalytic inquiry (Canestri and Glocer Fiorini 2009; Johnston 2005), concerning at least three different levels: *technical* (relating to the *duration* of the psychoanalytic relationship), *epistemological* (relating to the *evolution of* and *between* diseases or psychological stages), and *existential* (relating to the *consciousness* of the analyzing subject about himself). However, if subjective *well-being* is the ultimate purpose of the analytic experience, it is immediately clear that the last level defines also the previous ones. In this perspective, we can attribute an innovative role to Lacan's teaching: in receiving the Freudian psychoanalytical legacy as well as in participating in the contemporary thinking about subjective time consciousness, he tries to define *a proper scientific understanding of time* for psychoanalysis.

So, *where and how does Lacan address his inquiry about time?* In order to answer this question, it will be necessary to start from two preliminary observations. At a first *communicative* level, we must note that Lacan's knowledge transfer is basically "oral", since this was the form employed in his Seminar, and therefore was primarily destined to the formation of psychoanalysts. This is the reason we prefer to speak of Lacan's "teaching" more than of his "work", and is mostly the reason we have received Lacan's heritage in two related but substantially distinguished ways: the transcriptions of the lessons taken down by the Seminar's participants (assigned by Lacan himself to Miller's editing and not yet entirely published) and a small but theoretically controlled production of Lacan's writings (collected in two publications with the simple title of *Écrits 1966* and *Autres Écrits*, Lacan 2001).

At a second *methodological* level, the distinction between *The Seminar* and *Écrits* leads us to locate the one as the place of Lacan's creative dissertation about Freud's work and the practice of psychoanalysis, while the other shows Lacan's establishment of his own doctrine. Thereby, if in *The Seminar* the topic of time appears to be contingently necessitated by his addressing the psychoanalytic practice (i.e., related to the phenomena and the mechanisms of "removal"), in the *Écrits* it is the proper object of a well-framed analysis, expounded in the 1945 essay "Logical Time and the Assertion of Anticipated Certainty" (Lacan 1966: 161–175). Thus, we might encounter the topic of temporality disseminated throughout the *Seminar*, but it is only by starting from the aforementioned essay that an otherwise contingent reflection may receive its coherent articulation.

Mainly, the aim of the essay is to undermine logic's pretensions of timelessness and eternity by showing how certain logical calculations include an inescapable reference to temporality. However, this is not specifiable by reference to the "clock time", but will be itself the product of certain logical articulations. *This distinction between logical and chronological time underpins Lacan's whole theory of temporality*, offering the French psychoanalyst an epistemological starting point from which articulate the existential and technical levels of his psychoanalysis. Therefore, in order to understand how this is made possible, we have to focus on the entire argumentation as Lacan proposes it in his essay, namely by discussing the so-called *sophism of the three prisoners.*

Once, a prison warden had summoned three choice prisoners and announced that:

> For reasons I need not make known to you now, gentlemen, I must free one of you. In order to decide which, I will entrust the outcome to a test that you will, I hope, agree to undergo.
>
> There are three of you present. I have here five disks differing only in color: three white and two black. Without letting you know which I will have chosen, I will fasten one of them to each of you between the shoulders, outside, that is, your direct visual field—indirect ways of getting a look at the disk also being excluded by the absence here of any means by which to see your own reflection.
>
> You will then be left at your leisure to consider your companions and their respective disks, without being allowed, of course, to communicate among yourselves the results of your inspection. Your own inter-est would, in any case, proscribe such communication, for the first to be able to deduce his own color will be the one to benefit from the discharging measure at my disposal (Lacan 1966: 161).

Since the prison warden fastens to each prisoner a white disk (therefore, not using any of the black disks), *how should the subjects solve the test*?

At this point, we (as the prisoners) have to keep in mind all the possible combinations generated by the presence of the five disks and then to consider the consequent possible occurrences (Žižek 1993: 73–74):

(1) *There are one white and two black disks.* If one prisoner has a white disk and the other two black ones, the one with the white disk can immediately "see"

that his own is white by way of a simple reasoning: "There are only two black disks; I see them on the others' shoulders, so mine is white".

(2) *There are two white and one black disk.* One of the prisoners sees one black and one white disk, so his own is either white or black. Then, he can suppose: "If mine is black, then the prisoner with the white disk would see two black disks and immediately conclude that his own is white (as in the previous case); since he does not do it, mine is also white".

(3) *There are three white disk.* This last possibility (that is evidently the one chosen by the warden) is also the most complex, since the reasoning goes here like this: "I see two white disks, so mine is either white or black. If mine is black, then any of the two remaining prisoners would reason the following way: 'I see a black and a white disk. So if mine is black, the prisoner with the white disk would see two black disks and would stand up and leave immediately. However, he does not do it. So mine is white. I shall stand up and leave'. But *since none of the other two prisoners stands up, mine is also white*".

Thereby, it is clear that last situation obtains its solution only with relation to the two previous ones: this consideration is necessary to understand the temporal dimension of the sophism as it is argumented by Lacan in his "dissertation". Here, Lacan outlines how an external observer (like the warden) could see that:

> […] after having contemplated one another for *a certain time,* the three subjects take *a few steps* together, passing side by side through the doorway: each of them then separately furnishes a similar response which can be expressed as follows:
>
> I am a white, and here is how I know it. Since my companions were whites, I thought that, had I been a black, each of them would have been able to infer the following: "If I too were a black, the other would have necessarily realized straight away that he was a white and would have left immediately; therefore I am not a black". And both would have left together, convinced they were whites. As they did nothing of the kind, I must be a white like them. At that, I made for the door to make my conclusion known.
>
> Thus, all three exit simultaneously, armed with the same reasons for concluding (Lacan 1966: 162).

According to Lacan's discussion of the sophism (Lacan 1966: 163–165), we can observe that this solution is not so "perfect", since it seems to introduce just one (temporal) *suspensive scansion*, while the sophism obtains its logical coherence by recognizing the value of two of them. In fact, each prisoner remains motionless twice: the first stop occurs when, at end of the reasoning, (every) one of the subjects observes that none of the other two stands up, and therefore, he deduces to be in the third situation (namely to have a white disk). However, since all of them would stand up simultaneously, (every) one might wonder if this happens because they all have really followed the same reasoning or rather because they are still in the first or second situation. So, it is only since (every) one stops (at least) for a second time that the hypothesis is finally confirmed and all of them can go out with the certainty of being white.

At this point, *we are already able to distinguish "chronological" from "logical" time*. Specifically, here we can make a first distinction between "cosmological" (or "natural") and "human" time. In fact, in the logical puzzle, a sequence of events occurs with a seemingly simple cause and effect relationship: the warden organizes the test; the prisoners proceed through the outlined reasonings; the prisoners gain their own freedom. But although these events seem to follow each other chronologically, nevertheless we can call into question the notion of a linear causality, since "the *interruption* [*temps d'arret*]" marked by the suspensive motions determines the chains of events as "signifying" (Lacan 1966: 166).

Thereby, it becomes clearer that by introducing and discussing sophisms, Lacan underlines how they bring into play several aspects pertaining to the "sensed action" rather than the simple "physical movement": doubt, haste, reasoning, choice, in a few words, *the time of the subject's consciousness*. In this perspective, starting from the two suspensive scansions, we can see the sophism as temporally modulated by three "evidential moments" (Lacan 1966: 167–169), each of them supported by a different kind of subject:

(1) *The instant of the glance (or the look)—the impersonal subject.* On finding himself in front of two blacks, each subject effects the same *logical exclusion*: he knows that he is white. But each subject sees two whites, so this first possibility (two blacks and a white) is excluded for each subject. Thus, at first glance, what is important is what *each subject does not see,* that is, the excluded black disks. Lacan calls this *"initial evidence"* or *"the given of the problem"* (Lacan 1966: 165, 167) and he connects it to *an original interdiction*, that is, the combination of three blacks. This is a time "without duration", which does not pass, the components of which are identical to themselves. All of the knowledge, or rather, the non-knowledge that is received in this initial time, must be formulated on the level of *the impersonal pronoun* "one". Lacan gives the phrase "one knows that…" (Lacan 1966: 167) and adds that this phrase is related to the real unknown of the problem, which is the *ignored attribute of the subject himself.*
(2) *The time for comprehending—the indefinite reciprocal subject.* "Were I a black, the two whites that I see would waste no time realizing that they are whites": here some time had to elapse, a certain "time for understanding", that involves intersubjectivity; in fact, in order for me to arrive at the conclusion that my disk is white, I have to "transpose" myself into the other's reasoning (if the other with the white disk were to see a black disk between my shoulders, he would immediately know that his must be black and stand up; since he does not do it, mine is also white). The subject would remain paralyzed if he did not observe in the others the same experience of absence from which his desire for understanding began. Then, this kind of subjectivity remains that of the *indefinite reciprocal subject*, as Lacan puts it: a simple reciprocal capability of taking into account the other's reasoning.
(3) *The moment of concluding—the assertive subject.* "I hasten to declare myself a white, so that these whites, whom I consider in this way, do not precede me in recognizing themselves for what they are": here, (every) one concludes that he must be a white because he sees that the other subjects hesitate and, for this reason, they must be in front of two whites. This subject must now act in a hurry in order to beat out the other two subjects who must be coming to the

> same conclusion. It is only this third moment which provides "the true 'genesis' of the I": from "the void of the subject epitomized by the radical uncertainty as to what I am", to "the conclusion that I am white, to the assumption of the symbolic identity—That's me!" (Žižek 1993: 75–6); here comes the *urgency of the assertive subject*.

However, in order to be able to recognize the assertive subject in the other places of Lacan's teaching, it is necessary to notice that *this conclusive element of assertive haste is brought about not by the prisoners' eagerness to escape imprisonment but by their logical position*. Indeed, the decision that the prisoners come to regarding the color of their respective disks is entirely based on the others' pausing for a second time. Thus, if the others cease their second pause and move on, the whole basis of the certainty will be destroyed. Lacan calls this anticipation of certitude the fundamental form of a *collective logic* (Lacan 1966: 174). Then we can observe that (every) one achieves his certitude only retroactively, once the end is reached; but at the same time, this is an "announced" certitude, since it is related to the temporal tension of a logical necessity.

Finally, we have reached the point where Lacan's inquiry on time—as it relates to the topic of subjectivity—shows its own peculiarities and encounters its own difficulties. In fact, the argumentation makes it possible to distinguish "logical" time from both "natural" time and time as subject's feeling, but at the price of losing every possible reference to a phenomenological self-grounding "ego". In other words, the subject of psychoanalysis is not that of the philosophical "will", since as an impersonal subject it does not have an immediate consciousness of itself, but it has to rescue it retroactively, and then finds the moment of its beginning "negatively", a posteriori.

In this regard, the distinction made by Žižek (1993) between *subjectivity* and *subjectification* could be useful: since the prisoners do not see the disk between their shoulders, they initiate a series of efforts to give meaning to their experience. In other words, the subjects do not already know "who they are" and so they initiate a research of subjectification (that which in Lacan's teaching will constitute the ground of the Subject's desire) in an attempt to reabsorb the void revealed by the original interdiction (the impersonal subject) to subjectivity (the assertive subject). The assertive subjectivity finally manifests itself as a judgment in an act, and its time of subjectification is what—in his 1960 essay "Position of the Unconscious"—Lacan will call *reversible time* (Lacan 1966: 711), identifying its core in the tension between *anticipation* and *retroaction*.

Après coup (*retroaction*) is the term used by Lacan (Lacan 1966: 281, S11: 197) to translate Freud's *Nachträglichkeit* (which the *Standard Edition* renders as "deferred action"; Evans 1996: 209). These terms refer to the way present events affect past events a posteriori in subjectification: what concerns psychoanalysis is not the chronological past sequence of events in themselves, but the way the patient retroactively signifies ("rewrites") them. Then, if retroaction refers to the way the present affects the past, anticipation refers to the way the future affects the present (the time of concluding is reached on the basis of the time of comprehending). According to Lacan, the structure of anticipation is best illustrated linguistically by the *future perfect tense* (Lacan 1966: 306): in the future, the past "will have been"; or, in other words, we can say that the tense between après coup and anticipation is a "split between 'I don't know yet' and 'oh yes, I already knew that'" (Soler 1996: 64).

Starting from this, and unlike the other forms of psychoanalysis based on a linear concept of time (i.e., Ferenczi's perspective), in Lacan's teaching it is impossible to define a linear sequence of developmental stages through which the subjectification naturally passes. This is also how he resignifies some of the most well-knowned Freudian pages; for instance, in "the three times of the Oedipus complex", where the ordering is one of the logical priorities rather than of chronological sequence; or in the reworked analysis of narcissism as a "mirror stage" (Lacan 1966: 75–81; Lacan 1975: 74; Lacan 1981: 49–50; Lacan 1998: 121), where the "ego" is constructed on the basis of the anticipation of an imagined future wholeness (which, in fact, never arrives).

Thus, when Lacan argues that the aim of psychoanalytic treatment is "the complete reconstitution of the subject's history" (Lacan 1975: 12), he makes it clear that what he means by the term "history" is "less a matter of remembering than of rewriting history" (Lacan 1975: 14). Then, what concerns psychoanalytic treatment is not the "real" past sequence of events in themselves, but the way that these events exist "actually" in the patient's experience, posing questions of subjectivity, history and future, in terms of the knot of connection and resistance. For these same reasons, it is precisely in relation to the notion of the subject's "history" that we match one of the thorniest point of Lacan's teaching, since it affects the relationship between psychoanalysis and language.

In fact, it is clear at this point that *the time of subjectivity is the time of language*: a twofold temporality that comprises the anticipation of the end of the sentence (at which point the full meaning will be given), and the moment when the sentence ends and what was said retroactively takes on a meaning (Soler 1996: 61–66). Nevertheless, Lacan does not mean by this that we have to wait for the end of a sentence to garner any meaning from it, for we anticipate meaning as the sentence proceeds, the signifier positioning its dimension, its meaning, somehow ahead of itself: like retroaction, anticipation marks the structure of speech (Lacan 1966: 419).

Here, at the end of a remarkable transition from Saussure's linguistics of signs (2013) to Benveniste's linguistics of discourse (1971), Lacan highlights the role that *punctuation* plays in the formation of utterances. Punctuation consists of an act by which, using the interpunction marks, we segment the discourse in significative semantic modules: recall that at the end of the sentence, once the full stop is reached, a form of retroactive meaning is produced. In relation to this, Lacan states that "punctuation, once inserted, establishes meaning; changing the punctuation renews or upsets it; and incorrect punctuation distorts it" (Lacan 1966: 258). Thus, he can also observe that what pertains to the end of sentences can be considered in relation to the end of analytic sessions, and starting from this point, he also theorizes the use of *sessions of variable duration* (*séances scandées*).

In fact, from what has been said so far, we have to deduce that none can predict the time for concluding a session, and thus that it is impossible to decide "a priori" the session's duration. Particularly, this means that although the analyst can use punctuation to produce a cut in the analysand's discourse—and so to precipitate the concluding moment—nevertheless it remains an act immanent to the discursive structure itself. Thus, we have to acknowledge again (as we have already seen in the case of the assertive subject) that in Lacan's psychoanalysis the time of subjectivity is an *effect of texture*; in other words, if "the unconscious is structured as a language" (Lacan 1964: 20, 2009:40) then "the unconscious has no time" (Freud 1915).

However, the utterance/punctuation pairing soon shows its limits. In fact, it is clear that individuals give meaning to themselves by producing "texts", that is to say, by producing semantic units more complex than an utterance. Then, if the aim of psychoanalytic treatment is to help the patient "rewrite" them correctly, how can it guarantee at the same time a continuity in difference in the transfer from text to text? How can we prevent the risk of transforming psychoanalysis into a form of philosophy (namely, into a metalanguage that pretends to describe texts from the "outside")?

Thus, despite the results obtained using the lexicon of discoursive linguistics, during his teaching Lacan increasingly displays his need to formalize psychoanalysis (the use of algebraic symbols began in the 1950s: Lacan 1978a, b; Lacan 1966). Since the Seminar of 1964–1965 (Lacan 1965), the French psychoanalyst has connected the essay on logical time with his developing notions of the "topology of time", but it is only in his late teaching (Lacan 1977; Lacan 1978a, b; Lacan 1979) that he directly engages the epistemological level of time analysis, trying to offer constants to the dynamic of time consciousness, using *topological models* (Johnston 2005; Ragland-Sullivan and Milovanovic 2004; Ragland 2015).

Topology is "a branch of mathematics which deals with the properties of figures in space which are preserved under all continuous deformations" (Rabaté 2003: 210). Lacan argues that it is not simply a metaphorical way of expressing the concept of structure: *it is structure itself.* He emphasizes that topology privileges the function of the *cut (coupure),* since the cut is what distinguishes a discontinuous transformation from a continuous one. In other words, in the cut, change is not seen as a gradual or smooth move along a continuum, but as an abrupt shift from one discrete structure to another: as the cut operates a discontinuous transformation in the topological figures, so an effective interpretation modifies the structure of subject discourse in a radical way (see Figs. 1 and 2).

Finally, therefore, the function of the cut allows us to reread the three temporal movements of subjectivation exposed in the sophism, understanding them as the timeless performance of a musical score. Then, in a way that anticipates some of the current most validated research in semiotics (Greimas) and mathematics (Thom), but that remains only in a "sketched" form due the death of the French psychoanalyst, the last word of Lacan on Subject's time is "Rhythm".

Fig. 1 In this Möbius strip, he cut operates a passage from the projective plane to the trefoil knot

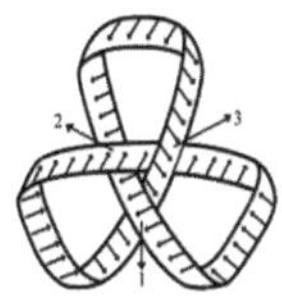

Fig. 2 Here we see the three singular points that the duplication transforms into crossing points. The knot's possibility or impossibility of intersecting itself in a double point, a passableness that involves a moment of indecision—neither yes nor no—inscribes the knot in the field of ternary logic. This property of knots necessitates a space of immersion of more than three or even four dimensions, since the topological flexibility of the knot entails the fourth dimension. The immersion of the knot in three-dimensional space projects the temporal dimension across an expanse as a succession (Prieto 2011)

References

Detailed lists of Lacan's publications and Seminars are available in: Rabaté (2003: 272–281), and Roudinesco (1993: 511–534).

Benveniste, É. (1966). *Problèmes de linguistique générale* (Vol. 1). Paris: Gallimard. English edition: Benveniste, E. (1971). *Problems in General Linguistics* (M. E. Meek, Trans.). Coral Gabes: Miami University Press.

Canestri, J., & Glocer Fiorini, L. (Eds.). (2009). *The Experience of Time: Psychoanalytic Perspectives*. London: Karnac.

Dylan, E. (1996). *An Introductory Dictionary of Lacanian Psychoanalysis*. New York: Routledge.

Freud, S. (1915). Das Unbewusste. *Internationale Zeitschrift für Psychoanalyse* 3: 189–203. English edition: Freud, S. (1957). The unconscious. In: *The Standard Edition of the Complete Psychological Works of Sigmund Freud* (Vol. XIV, pp. 166–215) (1914–1916) (J. Strachey, Trans.). London: Hogarth Press.

Johnston, A. (2005). *Time Driven. Metapsychology and the Splitten of the Drive*. Evanstone (Ill): Northwestern University Press.

Lacan, J. (1964). *Le séminaire. Livre XI. Les quatre concepts fondamentaux de la psychanalyse* (1963–1964). Paris: Seuil. English edition: Lacan, J. (1979). *The Seminar. Book XI. The Four Fundamental Concepts of Psychoanalysis, 1963–64* (A. Sheridan, Trans.). London: Penguin.

Lacan, J. (1965). *The Seminar XII, Crucial Problems for Psychoanalysis, 1964–1965* (C. Gallagher, Trans.), Unpublished typescript.

Lacan, J. (1966). *Écrits*, Paris: Seuil. English edition: Lacan, J. (2006). English edition: *Ecrits*. (B. Fink et al., Trans.) New York: W.W. Norton and Company.

Lacan, J. (1973). L'étourdit. *Scilicet* 4 (pp. 5–52). Seuil: Paris. English edition: Lacan, J. (2009). *The Letter* 41 (pp. 31–80) (C. Callagher, Trans.) (now in Lacan 2001: 449–496).

Lacan, J. (1975). *Le séminaire. Livre I. Les écrits techniques de Freud (1953–1954)*. Paris: Seuil. English edition: Lacan, J. (1988). *The Seminar. Book I. Freud's Papers on Technique, 1953–54* (J. Forrester, Trans.). New York: Norton; Cambridge: Cambridge University Press.

Lacan, J. (1977). *Le séminaire. XXIV, L'insu que sait de l'une bévue s'aile a mourre, 1976–1977*. Unpublished typescript.

Lacan, J. (1978a). *Le séminaire. Livre II. Le moi dans la théorie de Freud et dans la technique de la psychanalyse (1954–1955)*. Paris: Seuil. English edition: Lacan, J. (1988b). *The Seminar. Book II. The Ego in Freud's Theory and in the Technique of Psychoanalysis, 1954–55* (S. Tomaselli, Trans.). Cambridge: Cambridge University Press.

Lacan, J. (1978b). *The Seminar XXV, The Moment to Conclude, 1977–1978* (C. Gallagher, Trans.), Unpublished typescript.

Lacan, J. (1979). *Le séminaire. XXVI. La topologie et le temps, 1978–1979*. Unpublished typescript.
Lacan, J. (1981). *Le séminaire. Livre III. Les psychoses (1955–1956)*. Paris: Seuil. English edition: Lacan, J. (1993). *The Seminar. Book III. The Psychoses*, 1955–56 (R. Grigg, Trans.), New York: Routledge/Norton.
Lacan, J. (1998). *Le séminaire. Livre V. Les formations de l'inconscient* (1957–1958). Paris: Seuil. English edition: Lacan, J., *The Seminar. Book V. The formations of the unconscious*, 1957–58. (C. Gallagher, Trans.). Unpublished typescript).
Lacan, J. (2001). *Autres écrits*. Paris: Seuil.
Prieto, G. (2011). Writing the subject's knot. *Recherches en psychanalyse, 12*(2), 170–179. http://www.cairn.info/revue-recherches-en-psychanalyse-2011-2-page-170.htm.
Rabaté, J.-M. (Ed.). (2003). *The Cambridge Companion to Lacan*. Cambridge: Cambridge University Press.
Ragland, E. (2015). *Jacques Lacan and the Logic of Structure: Topology and Language in Psychoanalysis*. New York: Routledge.
Ragland-Sullivan, E., & Milovanovic, D. (Eds.). (2004). *Lacan: Topologically Speaking*, Berkeley: Ragland Milovanovic.
Roudinesco, É. (1993). *Jacques Lacan. Esquisse d'une vie, histoire d'un système de pensée*, Paris: Fayard. English edition: Roudinesco, E. (1997). *Jacques Lacan* (B. Bray, Trans.). New York: Columbia University Press.
Saussure, F. (1916). *Course de linguistique générale*, Paris: Payot. English edition: Saussure, F. (2013). *Course in General Linguistics* (R. Harris, Trans.). London: Bloomsbury.
Soler, C. (1996). Time and Interpretation. In: F. Richard, F. Bruce & J. Maire (Eds.), *Reading Seminars I and II: Lacan's Return to Freud* (pp. 61–66). Albany: State University of New York Press.
Žižek, S. (1993). *Tarrying with the Negative. Kant, Hegel, and the Critique of Ideology*. Durham: Duke University Press.

Time, Narration, Memory: Paul Ricoeur's Theory of History

Giuseppe Cacciatore

Abstract The theme of the historical experience of the finite man is what allows Paul Ricoeur to complete a long journey that, from the original agreement with a strictly eidetic phenomenology—through the analysis of the will and its sensible and corporeal instincts—leads him to a life's hermeneutics that is firstly the understanding of "ontological deficiency", as the basic trait of the human will's being, of its passions, of its fallibility and continuous exposure to guilt. But Ricoeurian hermeneutics starts from the refusal of every abstract absolutism of the spirit and of its forms, as well as of a similarly abstract idea of the universal essence of the human. And it's along this process that the further moving of perspective occurs towards the hermeneutics of a text, that becomes objective in the story and in its writing and, even more, in the world and in its stories.

I have already argued in other pages I have written on Ricoeur (Cacciatore 2013) that his intellectual trajectory can be identified with a long process of philosophical clarification of the finite, fallible and historical character of human will. From the original interest in phenomenology and the philosophy of existence until the mature results of his original hermeneutical perspective (Ricoeur 1995), what progressively matures are the foundational elements of a critical theory of history that is also, and fundamentally, research into a renewed nexus between time and life. "La philosophie réflexive avait opposée le transcendental à l'empirique, dans l'idée que l'historique n'était qu'un aspect de l'empirique; le transcendental pouvait alors être tenu pour anhistorique. Voici maintenant que l'historique entrait dans le champ réflexif, changeant à la fois le sens du transcendental et celui de l'empirique" (Ricoeur 1991). Along this path there is increasingly delineated the necessity of the link between *self-understanding* and history: "une compréhension de soi qui était la reprise réflexive de sa propre histoire, une histoire qui offrait à la réflexion pure la méditation longue qui manquait au court-circuit du Cogito cartésien se pensant

G. Cacciatore (✉)
Philosophy Section, Department of Humanities, University of Naples Federico II, via Porta di Massa 1, 80133 Naples, Italy
e-mail: giusepca@tiscali.it

F. Santoianni (ed.), *The Concept of Time in Early Twentieth-Century Philosophy*, Studies in Applied Philosophy, Epistemology and Rational Ethics 24,
DOI 10.1007/978-3-319-24895-0_19

lui-même" (Ricoeur 1991). The dialectics of the voluntary and involuntary, of rational representation and lived experience, of universality of consciousness and contingency of the limits of finite man, innervates a new vision of time that no longer finds its foundation in the formal a priori of reason, but in the concrete and always new relation that human existence constructs with ethical, political and anthropological expressions of past history and its future projection.[1]

The topic of history, of the historical experience of finite man, is the trace that insinuates itself within the entire *corpus* of Ricoeur's theoretical edifice: from the originary adhesion to a rigorously eidetic phenomenology, that is founded on the analysis of the will and its sensible and corporeal pulsions, to the hermeneutics of life that is above all comprehension of the "ontological deficiency" as a founding trait of the being of the human will, of its passions, of its fallability (Ricoeur 1986a: 80). But Ricoeurian hermeneutics moves from the rejection of all abstract absolutization of the spirit and its forms, as well as of an equally abstract idea of universal essence of the human. All his works, especially those starting from the early 1960s, appear thus pervaded with a continuous search—above all on the concrete anthropological and historical level—for the possible points of mediation of reason and passion, voluntary and involuntary, finiteness and desire for transcendence, aspiration for totality and satisfaction of the material needs of existence. The dialectical (of an open dialectics and without conclusive summaries) of the three fundamental movements of the human being—the transcendental activity of reason, the sentimental sphere of the affections, the practical activity of the single individuals[2]—constitutes the necessary bridge over which to pass to a *historical-hermeneutical* phenomenology that utilizes, on the one hand, the function, both synthetic-transcendental and memorative, of the imagination (an instrument, for that matter, that is indispensable for mediating between the finiteness of historical-practical experience and the generalization of symbols and myths) and, on the other, the conviction that there is an originary constitution of the finite subject as the place of the conflict of interpretations and as the ever new possibility of creation and imagination of symbols and myths. The historical-ethical superstructure of Ricoeur's hermeneutic phenomenology passes mainly through the conviction of how much the narrative contents and forms of the myths and symbols count in the rethinking of a philospohy that finds its bases in the concreteness of language and of its manifestations and in the determination of an anthroppological vision that, with the courageous rivindication of a responsable subject, the creator of his world of signs and reflections, managed to move about in an autonomous and often critical manner at the height of structuralist flourishing. It is along this ever more complex process, rich in articulations in the field of history and of

[1]Naturally the reference is to the great Ricoeurian works in which he faces, among other things, the problem of history (Ricoeur 1955, 1983, 1984, 1985, 2000).

[2]There seem to me evident here some elements that render the Ricoeurian discourse similar to the historical research of Dilthey and his idea of *entirety* of the objectizations of the human individual on the level of reason, of the sentiment, of action. On Ricoeur's judgement of Dilthey one may consult the pages of the last part of *La mémoire, l'histoire, l'oubli*.

psychoanalysis, in that of literature and of religion, of sociology and of anthropology, of linguistics and of semiology and, naturally, in that of philosophy, that there occurs the further shift in perspective towards a hermeneutic of the text, of a text that becomes objective in the telling and in its writing, and even more, in the world and its stories (Ricoeur 1984, 1985, 1986[2]).

The French debate, and not only that, on the theory and history of consciousness (and science) of history—revived again between the first and second half of the Twentieth century by the school of the *Annales*—is enriched with the contributions made by Ricoeur on metaphor and on the relation between temporality and narration. It was a reflection that aimed essentially at clarifying a philosophical idea of narrative, semantic and symbolic-imaginative creativity of the individual[3]; a creativity that, at the same time, allows the historian and his narration to recreate (according to a practice that is close to the Diltheyan *Nacherleben*) the very conditions of reproduceability (or of *Mimesis*, as Ricoeur 2000) of the temporal lived experience. All this is expressed limpidly in the enunciation that opens the third chapter of the first volume of *Temps et récit*. "Il existe entre l'activité de raconter une histoire et le caractère temporel de l'expérience humaine une corrélation qui n'est pas purement accidentelle, mai présente une forme de necessité transculturelle [testifying how strong the link is that Ricoeur establishes between his idea of historical narrative and the ethical-practical dimension]. Ou, pour le dire autrement: *que le temps devient temps humain dans la mesure où il est articulé sur un mode narratif, et que le récit atteint sa signification plénière quand il devient une condition de l'existence temporelle*" (Ricoeur 1983: 105). What is at play—as Ricoeur clarifies well in the conclusions of the third volume of *Temps et récit*—is the motifavion (and, with it, also the theoretical finality) that underlies the relation of what is defined, on the one hand, *the aporetic of time*, and, on the other, the *poetics of the tale*. In brief, the indefinite character of time, the mystery of time—Ricoeur writes—does not mean that it remains unthought or even linguistically unexpressable. For this reason, the "*reaffirmation of historical consciousness*", if it wishes to aspire to the truth, cannot do without entrusting itself to continual research, on the part of individuals and communities, for their respective narrative identity. It is in this crucial passage that there is shown what Ricoeur calls *the hard kernel* of his entire research: "the inseparable relation between *aporetic of time* and *poetics of the tale*". It thus becomes simple to understand the motives that underlie —for example in the third volume of *Temps et Récit*—an original idea of *narrative identity* (and one could evaluate how much this crucial theoretical and methodological passage of Ricoeur's might serve in founding a critical vision of multicultural contemporeneity), which also shows the ethical valence of the reflections

[3]This is not the place for doing so, but it would doubtless be interesting to see how many assonances can be found between a classically historical-philological and philosophical vision of metaphor, like that of Vico, and the Ricoeurian idea of metaphor as the critique and overcoming of a hermeneutics that is only methodical and analytical-lingusitic. Like Vico, Ricoeur also speaks of the metaphorical construction as of a poem in minature, capable of transfiguring and creating, through fantasy and the imagination, the same reality (Ricoeur 1975).

and of the analyses concerning the question of the other and of his recognition,[4] but also the deepenings that the last Ricoeur wished significantly to dedicate to the relation between history, memory and oblivion.

It appears clear, at this point, that the history-memory relationship does not halt at the phenomenological level, but passes over onto the epistemological level, and also the ethical-pragmatic level. But the clarification of the necessary and obligatory nexus between memory and history, needs to be referred to an "ontological" dimension, so to speak, of memory (and of its contents of temporalization), in the sense that without memory it is impossible to activate any historigraphical praxis. But Ricoeur knows well that the historical science could be exercised also independently of memory, in all those forms, for example, in which the contents of the historical tale are already given in the objectivized memories of other tales or of other places of conservation of the traces and of documentations, or, further still, in those long-lasting forms of historiography where memory and the telling become diluted, almost to the point of disappearing, in the dilation of the historical times of the human and physical-natural world.[5]

I retain, then, that the fundamental motives of Ricoeur's hermeneutical and historical-critical philosophy can certainly constitute the basis of a *theory of history for our times.* That which, in fact, a renewed function of historical reflection has need of is the affirmation of a strong distancing from every form of historical knowledge (and of its public use) that wants to present itself in forms and contents of absoluteness and of unicity and that pretends, in this way, to constitute itself as the collective subject of history. A theory of history for the contemporary world cannot but be "ontologically" marked by the pluralism, not only of points of view and of understandings of history, but also and above all of collective memories. There is another motive that authorizes us to consider the theory of history elaborated by Ricoeur as a useful instrument for the comprehension of modernity and of its transfiguring itself into post-modernity. I refer to the fact that for Ricoeur, there remains valid the Nietzschean and post-Nietzschean critique of the excess of historical culture; so that there always emerges again that "inquieting extraneousness of history" of which Ricoeur speaks, not by chance, in the pages dedicated to the relation between history and time. In short, taking up again Plato's *Phaedro* (and

---

[4]It is not by chance that Ricoeur dedicated his last efforts to precisely this topic (Ricoeur 2004). It is, moreover, an exemplary book for understanding the manner of proceeding of Ricoeurian philosophizing, all ennerved with theoretical problems that are encountered and discussed, however, along the paths of an original re-reading of genetical places (archaeological, so to say, but also lexicographical) of the word-key being researched. On account of which, without forced superimpositions of history and theory, one passes easily from the interpretations of the pages of Homer, Sophocles, Aristotle, Descartes, Kant, Hobbes, Hegel, Bergson, to the mediated hypotheses on the nexus between recognition and gratitude, recognition and identity, recognition and memory.

[5]Of the experience of the *Annales*, Ricoeur speaks at length, regarding the history of mentalities and of the first generation of the journal, that of the founders Lucien Febvre and Marc Bloch. Ricoeur addresses also the second generation that concentrates particularly on the relation between long duration and macrohistory (Ricoeur 2000).

perhaps not forgetting the metaphor used by Troelsch and by Dilthey on "Odin's lance" which, like history, wounds and at the same time heals the wounds), for Ricoeur, there still remains open the question whether the *pharmakon* of writing (and even more of historical writing) is a remedy or a poison. Thus the "disquieting extraneousness of history" can be understood, and perhaps even exorcised if one passes from the epistemological level (and I would add, certainly against Ricoeur, also from the ontological level) to the ethical level of the individual responsibility of the user of history. The reader of historical things, Ricoeur warns, must be and maintain himself above all as a *warned citizen*, capable of placing himself, on the level of public discussion, as a balancing element between history and memory.

I must finally give the reasons for my interpretive choice which prefers more the critical character of Ricocurian hermeneutics than the ontological one. This naturally, in my opinion, is a result of the centrality gradually taken on, in the work of Ricoeur, by the topic of historicity, of temporality, of memory, with all the features that this choice of method and of content have had on the ethical arc of the last phase of speculation of the French philosopher. But it is not only a question of this. What emerges fully is the appearance of a key of access which, more than others, allows one to recuperate and justify an ethical function of history,[6] or, to express it better, of the *historia rerum gestarum*, which is never completely exhausted, under pain of the destruction and uselessness of memory and the insignificance of oblivion, in the *Geschichte* of the *res gestae.* And all this is possible in the measure in which, in a virtuous relation with memory, this function of history places in contact the subject who writes history—the central theme of narration—with the historical subjects who have lived it. In this sense memory becomes active and alive and not only the place for gathering objective things to catalogue and explain, though taking into account the risk of the ideological and propagandistic use of historical memory. And yet, Ricoeur warns, we must never undervalue that in every fold of the historical life of individuals and of collectivities there hides the possibility of oblivion that not only can, in each moment, attack memory, but also have negative reverberations, to the point of calling it in question, on the pretext of truth or of objectivity of the historical representation. The theoretical and ethical purpose underlying Ricoeur's reflection is a complex, and at the same time, articulate vision of memory, which may hold together the cognitive and the psychic, the sensible-corporeal (the great phenomenological topic of *one's own body*) and the intentional lived experience, the normative and the affective, the neurophysiological and the historical-cultural, fleeing in this way from every mechanistically biological or idealistically philosophical reductionism. Therefore, both memory and oblivion belong, first of all, to the sphere of the lived experience, and are characterized by the capacity to remember or not the past, to relive it, but also to forget it, or to leave it silent until it is solicited to return to life. But the lived experience as the material

[6]Also in the North American philosophical context there has been addressed the topic of the ethics of history. A witness to this is Carr et al. 2004. Among the others, particularly noteworthy are the essays of Ankersmith, Danto, Margolis, Rüsen, and Makkreel (Ankersmith 2004; Danto 2004; Rüsen 2004; Makkreel 2004).

of the memory or oblivion of the past, represents the basis for their essentially pragmatic constitution, that is, for a phenomenology of the different means of manifestation of memory and of oblivion in the activity of the single man and of the human collectivities. I have elsewhere manifested my preference for the Ricoeur as philosopher, critic of history and theoretician of a hermeneutics of the historical condition and of its plural representations, rather than for the Ricoeur still tending towards an ontological and salvific dimension, represented by the still noble ethics of forgiveness. In short, for the Ricoeur who has harshly criticized every predefined hystory, every finalistic and absolutizing representation of it and has given, instead, space to a critical idea of historical knowledge, conscious of the limits of memory and of its expressions and manifestations, but not for this reason discouraged in the search for a *happy memory*. One can, in conclusion, affirm that, beyond the interweaving of the theoretical motives that range from phenomenology to hermeneutics, from epistemology to ontology, there emerges in the French philosopher a unitary thread, that of history and its representation, of history and its relationship with time, narration, and memory. In brief, that which in the end, in my opinion, constitutes the greatness and importance of the philosophy of Ricoeur remains fundamentally the conviction of the close relationship between the need for truth of historiography, on the one hand, and of narration, on the other. And this is made possible by the common denominator constituted by the temporal characteristic of human experience. "Le monde déployé par toute oeuvre narrative est toujours un monde temporel (…). Le temps devient temps humaine dans la mesure où il est articulé de manière narrative; en retour le récit est significatif dans la mesure où il dessine les traits de l'expérience temporelle" (Ricoeur 1983: 17).

References

Ankersmith, F.R. (2004). In praise of subjectivity. In: D. Carr, T. R. Flynn & R. Makkreel (Eds.) *The Ethics of History* (pp. 3–27). Evaston (Ill.): Northwestern University Press.

Cacciatore, G. (2013). La filosofia critica della storia di Ricoeur: narrazione, tempo, memoria. *Atti dell'Accademia Nazionale dei Lincei, classe di Scienze morali, storiche e filologiche. Rendiconti, 23*, 51–81.

Carr, D., Flynn, T. R., Makkreel, R. (Eds.). (2004). *The Ethics of History*. Evaston (Ill.): Northwestern University Press.

Danto, A. C. (2004). Prudence, history, time and truth. In: D. Carr, T. R. Flynn & R. Makkreel (Eds.), *The Ethics of History* (pp. 76–88). Evaston (Ill.): Northwestern University Press.

Makkrell, R. (2004). An ethically responsive hermeneutics of history. In: D. Carr, T. R. Flynn & R. Makkreel (Eds.), *The Ethics of History* (pp. 214–229). Evaston (Ill.): Northwestern University Press.

Ricoeur, P. (1955). *Histoire et verité*. Paris: Seuil.

Ricoeur, P. (1975). *La métaphore vive*. Paris: Seuil.

Ricoeur, P. (1983). *Temps et récit. Tome I: l'intrigue et le récit historique*. Paris: Seuil.

Ricoeur, P. (1984). *Temps et récit. Tome II: la configuration dans le récit de fiction*. Paris: Seuil.

Ricoeur, Paul. (1985). *Temps et récit. Tome III: le temps raconté*. Paris: Seuil.

Ricoeur, P. (1986a). *A l'école de la phénomenologie*. Paris: Vrin.

Ricoeur, P. (1986b). *Du texte à l'action*. Essais d'herméneutique II Paris: Seuil.

Ricoeur, P. (1991). Autocompréhension et histoire. In: T. C. Martines & R. A. Crespo (Eds.) *Paul Ricoeur. Los caminos de la interpretación*. Barcelona: Anthropos.
Ricoeur, Paul. (1995). *Réflexion faite. Autobiographie intellectuelle*. Paris: Éditions Esprit.
Ricoeur, Paul. (2000). *La memoire, l'histoire, l'oubli*. Paris: Seuil.
Ricoeur, P. (2004). *Parcours de la reconnaisance*. Paris: Stock.
Rüsen, J. (2004). Responsibility and irresponsibility in historical studies: a critical consideration of the ethical dimension in the historian's work. In: D. Carr, T.R. Flynn & R. Makkreel (Eds.), *The Ethics of History* (pp. 194–213). Evaston (Ill.): Northwestern University Press.

Historical Heterochronies: Evenemential Time and Epistemic Time in Michel Foucault

Agostino Cera

Abstract Proposing to examine syntheses of manifold experiences of the contemporary philosophical panorama, Michel Foucault's "critical ontology of actuality" culminates in the elaboration of an *epistemology of the human sciences* starting from their irreversible modern twist. Among the various possible ways of characterizing this epistemology—equipped with its own *modus operandi*: the archaeological-genealogical method—is to see it as the result of a reflection on the topic of temporality. In particular, it is a reflection on historical temporality as «knowledge of time», that is organized into two different yet complementary modalities, two *historical heterochronies* that can be defined as *evenemential time* and *epistemic time*. The first, acephalous and atelic, proves discontinuous, traced back to pure becoming. The second corresponds to the necessary disciplining of this original shapeless material by means of the solidification of structures that each time produce an equilibrium (an *episteme*) among the instances of knowledge and power. Yet these structures, too, remain subjected to that historic-temporal change to which they have attributed an order. Such an interpretation of time implies a metamorphosis of the notion of *subject* that, within an overall anti-humanistic perspective, goes from its death as the *cogito* to its rebirth in the form of an *ethical-aesthetic subject*: no longer the atemporal guarantor of the order of the real, but rather something committed to a risky exercise of its own freedom as *care of the Self*.

The treatment of time in Foucault constitutes a partial interpretative gamble. A question never explicitly thematized by the French philosopher, one may nevertheless consider it a fundamental *unthought* of his reflection: it is at once one of the main assumptions and one of the most important legacies of his "critical ontology of actuality".[1] Assuming such an awareness as a premise, the following

[1]On the concept of ontology of actuality—that survey that questions the current field of possible experience—see a short but dense essay of the last Foucault (Foucault 1984[3]: 32–50).

A. Cera (✉)
Philosophy Section, Department of Humanities, University of Naples Federico II,
via Porta di Massa 1, 80133 Naples, Italy
e-mail: moonwatch1@libero.it

F. Santoianni (ed.), *The Concept of Time in Early Twentieth-Century Philosophy*,
Studies in Applied Philosophy, Epistemology and Rational Ethics 24,
DOI 10.1007/978-3-319-24895-0_20

pages are intended to sketch a schematic reconstruction of this issue within the Foucaultian reflection, hoping thereby to help bring out its authentic profile. To turn this «unthought» into a hoped-for «to be thought about».

The work of Michel Foucault (1926–1984) has now risen to the rank of a classic, credited as a crucial step in the *redde rationem* that during the twentieth century philosophy has implemented against its own tradition. An unstable interweaving and a problematic synthesis of multifarious experiences of the contemporary philosophical landscape—from French classical epistemology (Bachelard and Canguilhem especially) to the destruction of metaphysics (Nietzsche and Heidegger); from neo-Marxism (Althusserian) to structuralism (Lévi-Strauss in the first place); from neo-Freudian psychoanalysis (Lacan) to the pragmatism of the linguistic turn (Wittgenstein)—Foucalt's contribution assumes, almost in spite of itself, the connotations of an *epistemology of human sciences* starting from their irreversible modern twist.

This "involuntary epistemology" was equipped primarily with its own *modus operandi*: the archaeological–genealogical method[2]. At the same time, it implies a carefully pondered choice of its own object, that it gleans precisely from the human sciences as disciplines "always animated by a sort of transcendental mobility" (Foucault 1966: 397): it involves apparatuses of knowledge that are never entirely normalizable in a scientific key and therefore the ideal litmus papers for that *positive unconscious of knowledge* that Foucault's analysis means to make explicit.

Among these constitutively unstable disciplines, pride of place belongs to history, «the knowledge of time» by definition. Foucault intends to reform historical knowledge from within, in particular liberating "the history of thought from transcendental subjection".[3] The «true historical sense» of which he makes himself the promoter—modelled on *wirkliche Historie*, the "effective history" theorized by Nietzsche in the *Second Untimely Meditation*—is that which "confirms our existence among countless lost events, without a landmark or a point of reference"

[2]As is known, the Foucaultian journey presents an internal evolution in terms of methodology. A first purely archaeological phase (whose important moments are: *The Order of Things*, 1966, and *The Archaeology of Knowledge*, 1969) takes as its object *episteme*, understood essentially as a discursive regime, as an archive. It is followed by a second (the turning point is the essay: *Nietzsche, Genealogy, History*, 1971), where his approach to the Nietzschean genealogy corresponds to his interest for the *device* (an expansion of the concept of episteme, which for its part can be defined as a specifically discursive device) and the centrality acquired by the theme of power, culminating in the world-famous analysis devoted to biopolitics.

[3]"My aim was to analyse this history, in the discontinuity that no teleology would reduce in advance; to map it in a dispersion that no pre-established horizon would embrace; to allow it to be deployed in an anonymity on which no transcendental constitution would impose the form of the subject; to *open it up to a temporality that would not promise the return of any dawn*. My aim was to cleanse it of all transcendental narcissism; it had to be freed from that circle of the lost origin, and rediscovered where it was imprisoned" (Foucault 1969: 224–225, the italics are mine).

(Foucault 1971: 89). The realization of this authentic historical sense involves the need to overthrow the main «continuist» model (metaphysical and anthropological): that of memory. A model that mortifies the monuments of the past, transforming them into documents, extorting from them words and meanings that they do not really possess. It will, therefore, be a question of reversing such an inertia, of "trying to operate a decentring that leaves no privilege to any center".[4] To "construct a countermemory—a transformation of history into a totally different *form of time*" (Foucault 1971: 93. The italics are mine).[5]

Therefore, the reform of history implies *ipso facto* a revision of the «form of time», of its image. On the other hand, as we mentioned, this is never made an explicit theme in his work. Schematically, it is possible to isolate two main modes of this reformed historical temporality: an *evenemential time* and an *epistemic time*. Two distinct but complementary times, two—very loosely paraphrasing the Foucaultian lexicon—*historical heterochronies*. Both subject to a fundamental assumption. As is known, Foucault's attempt to free thinking from its own, most rooted *idola* (meaning, origin, purpose, substance, cogito…) results in a provocative antiphilosophical attitude—empiricist and positivist—that "questions the teleologies and totalizations" (Foucault 1969: 18), rejecting any absolute overhistorical, understood as a salvific appeal to atemporal or extra-temporal entities. Well, this sacrifice of all forms of totalization, culminating in the affirmation of the "positivity of becoming" (Veyne 2008: 39)[6] involves its own necessary, albeit implicit, assumption of temporality as the only possible totality. *Everything is temporal.* Neither universals nor absolutes are given, nothing flees time and its fundamental device: historicity. Everything is immanent to time, which in turn is configured «essentially»—that is, in its most original way: *indeterminate* and *formless*—as a pure immanence.

This originary temporality is tantamount to the first image of time: the evenemential one. An acephalous and atelic time, lacking in any origin, direction or purpose, indeterminate and indeterminable except in the terms of the pure flowing of change and transformation. An agonic *apeiron* in which the *polemos* takes place between forces that, limiting and delimiting each other reciprocally, cause there to emerge, time after time, from their own interweaving, new combinations: the events. Differential singularities, atoms of discontinuity.

This evenemential incubator corresponds to the seminal, genetic horizon of historicity: it is history grasped in the act of its making. Evenemential time thus

[4]"The role of such a discourse […] does not set out to be a recollection of the original or a memory of the truth. On the contrary, its task is to *make* differences: to constitute them as objects, to analyse them, and to define their concept" (Foucalt 1969: 227).

[5]In this regard, Foucault had already affirmed: "If philosophy is memory or a return of the origin, what I am doing cannot, in any way, be regarded as philosophy; and if the history of thought consists in giving life to half-effaced figures, what I am doing is not history either." (Foucault 1969: 228).

[6]Later on Veyne further defines this positivity, as "nothing is negative, everything is positive, nothing is lacking" (162, note 20).

incarnates the *real time of history*. A widening, an undefined opening which is the theatre of the encounter/clash between elements pregnant with historicity.[7] Something not too unlike the clearing, the *Lichtung* described by Heidegger, but in comparison with which it lacks the ekstatic and above all aletheiological component. That is to say, that this immanent temporality does not cause *the* truth happen ontologically—it does not phenomenize the truth, interpreted as an event—but rather it generates within it the conditions of possibility (the empistemies, the devices) so that time after time "games of truth" are produced (Foucault 1988): operative and functional criteria, primarily practical-discursive rules and procedures, that serve as a glue for those same epistemic frames that have generated them.

Evenemential time represents the limit of archaeological–genealogical inquiry, whose penetration cannot go beyond the mere verification of the spectacle of this originary opening, of the complexive and multiform panorama of interweavings, mutations and successions of forces. A panorama in whose indeterminacy there stand out the compact bodies of the different evenemential singularities (again: the epistemies, the devices), each completed and defined in itself. Each incomparable to from the other, all irreduceable to a common horizon or a unitary and continuous weft. Individual postivities linked exclusively by differential nexuses.[8]

Evenemential temporality thus acts as a primordial time, behind and at the bottom of epistemic time, which instead corresponds to the rhythm marked within the historical singularities that have already emerged, to the rules of the temporal game in force within an already solidified historical positivity. If before we were in front of history in the making, here we are in the presence of a history already made, already told. That is, already regulated by an enunciative-discursive regime and by now normed and normalized. Unlike evenemential temporality, which captures historicity in real time, we are dealing with a *deferred temporality*, the recording of an event that has already happened. Epistemic time can thus find its worth as a kind of *historical consciousness*, the time of a historicity that turns back on its steps to rebuild them, capturing genealogically their rules of formation and

---

[7]Hazarding a consideration on a terrain that Foucault was not fond of frequenting, the image just described suggests an ontological layout of Heraclitan inspiration (according to the famous fragment B 53) even more than Nietzschean, since this idea of the primacy of contention underlies not so much a will to power, as a fundamental modality of relation. As a result, power as «the mothertonge of the *polemos*» (in its turn, the originary link between things) should not be considered and interpreted as a destructive element, but a producing and productive force.

[8]To make the schematization proposed in these pages more concrete, let us take as an example the general layout of *The Order of Things*. The individual examinations that read from within the devices in force in the Renaissance, the classical age and modernity belong to an epistemic temporality, while the horizon that looks panoramically, synthetically and simultaneously at the totality (discontinuous and differential) of such devices is evenemential.

operation: not the "origin" (*Ursprung*), then, but the "descent" (*Herkunft*) and "emergence" (*Entstehung*).[9] Obviously, such a hypothetical historical consciousness must be thought of in Foucaultian terms: anonymous, non-rational, non-projectual...

Epistemic time as the internal time in each of these compact spheres of differential singularities (events) is the point of application of conventional historical knowledge—the space of the origin, of continuity and of memory—as well as the natural ground of archaeological–genealogical survey, which aims to reveal the gears of the mechanism of a device, but in a rigorously formal, neutral and superficial key. Abstaining, that is, from any undue assumptions of content: the criterion by which we must decipher an epistemic regime is neither the meaning nor the purpose. "*La storia non ha senso*" ("history does not make sense"; Foucault 1977: 9),[10] its grip is given only by the action of regulatory strategies, with disciplining games of truth which, in turn, due to functional needs produce devices: logical, axiological, ethical, political, aesthetic... Now therefore, the task of authentic historical knowledge is to describe, to ascertain. Renouncing further claims, which would be nothing but relapses into transcendental narcissism.

The temporal coherence existing within a single epistemic regime—that is to say: the temporality peculiar to a specific historical period—corresponds to the densifying, the coagulating of a series of discontinuous and different durations around an ordering centre of gravity that acts literally as a metronome, and marking and imposing a unique rhythm on these durations. Thus making them co-present and compossible, compacting them around pillars within whose perimeter common rules are in force: the same games of truth and the same practices of knowledge and power. To give a concrete example: the role of a centre of gravity is played by similitude in the Renaissance, by representation in the classical age and by man (by the "analytic of finitude") in modern times, according to the archaeological reconstruction presented in *The Order of Things*.

Therefore, progressivity and linearity, while perceptible in epistemic time, do not at all serve as signs referring to a possible *ratio* of history, a *ratio* which, from within or without, drives its course. Rather, they constitute the effect and the proof of the grip of the centre of gravity of a particular device, of its correct functioning.

---

[9]Faithful to the Nietzschean model, which he even radicalizes, Foucault's genealogy makes the teaching of history its own, according to which things "have no essence". In this way, the myth of the *origin* is substituted by the concrete elements of the *descent* ("to maintain passing events in their proper dispersion") and of the *emergence* ("the principle and the singular law of an apparition"). On this topic see Foucault 1971: 78, 81, 83.

[10]The model of authentic historical knowledge is that "*della guerra e della battaglia. La storicità che ci trascina e ci determina è bellicosa [...] la relazione di potere, non la relazione di senso. La storia non ha «senso»; il che non vuol dire che sia assurda o incoerente. Essa è al contrario intelligibile [...] ma secondo l'intelligibilità delle lotte, delle strategie e delle tattiche*" ("of war and of battle. The historicity that drags us along and determines us is bellicose [...] the relation of power, not the relation of meaning. History has no «meaning»; which does not mean that it is absurd or incoherent. It is on the contrary intelligible [...] but according to the intelligibility of the fight, of strategies and of tactics") (Foucault 1977: 9).

The demonstration that in that given context, an order—i.e.: a balance and a hierarchy of forces, the outcome of a preliminary contention of theirs—has been established and is in force.

The interpretation of temporality just illustrated, and that in any case carries with it some aporetic traits, cannot be valid for the entire philosophical parable of Foucault, as it undergoes a significant change in the course of his work. Closely tied to the transformation sustained by another main theme (this time quite explicit) of his reflection: the *subject*. It is a natural connection, since the subject represents the main outcome of each epistemic regime, the most significant result of the processes that pass through it, the point of intersection and synthesis between its apparatus of knowledge and power. As if to say: each device or episteme produces both a specific subjectivity and a peculiar temporality.

Therefore, the notion of subject in Foucault encounters a substantial mutation. A metamorphosis that, within a perspective that is in any case antihumanist, goes from its death—in the classic form of a theoretical I: a *cogito*, according to the classical Cartesian and then Husserlian paradigm—to its rebirth, in the form of an ethical subject: no longer a timeless guarantor of the order of reality, but its actor, engaged in a risky exercise of one's Self as a "practice of freedom" (Becker 1987). Therefore: *from the cogito to the I, from the I to the Self.*

More specifically, the drastic refusal of any anthropologism, to which the deconstructive project (archaeological and genealogical) corresponds in regard to modernity, is matched by a radical critique of metahistorical and monodirectional history. "Continuous history is the indispensable correlative of the founding function of the subject" (Foucault 1969: 14), the myth of continuity is equivalent to the real "anthropological sleep", to the historical disease from which the human sciences must be liberated. The only liberating gesture is to recognize and apply the value of discontinuity, difference and singularity, definitively decreeing the death of man, and with it the extinction of the whole modern paradigm.

The suppression of anthropological idols (cogito and I) does not imply the final farewell of the idea of the subject, but its fundamental revision. As I said, once dead, man can be reborn in a new form. Less and less a product of subjection, increasing the result of an authentic subjectification. No longer reason, no longer conscience, the subject is rehabilitated in an ethical key. Ethical-aesthetic, to be precise: it is as a possible creator of an *aesthetics of existence* that it again becomes central in the last phase of Foucault's work, particularly with the second and third volume of the *History of Sexuality* and the last lectures at the Collège de France (Foucault 1984, 1984^2, 2001).

On the basis of this review, it is possible to note a subsequent reworking of the question of time. Having set aside the iconoclastic phase against traditional historical knowledge, there seems to take place a partial (and unspoken) rehabilitation of the continuist approach: precisely within an ethical-aesthetic horizon, the radical discontinuity underlying evenemential temporality is tempered in favour of a more conventionally historical conception and one witnesses the partial rehabilitation of a linear temporality. More generally, the two historical heterochronies outlined above limit their respective distance in the name of a continuity that legitimizes a common summons, not slavish but productive, to the past in the name of the urgencies of the present.[11]

One thus discovers that the new, hoped for subjectivity is not a historical novelty. The *Care of the Self* of late antiquity (the Platonic *epimeleia heautou*), the building up of oneself and according to the dictates of a style of one's own subjectivity as opposed to the Christian pastoral (a subjectification hetero-directed by a code of obedience to rules), becomes a clear trace of a new image of time. A smoother linearity which, without suggesting any design/project at its foundation —according to Foucault, history remains the realm of the «senseless and without purpose»—leaves recognizable traces of its journey. It points out trails that are passable again, by means of an ontology of actuality and of a reformed subject that possesses the keen gaze (the fineness of historical sense) to rediscover them.

References

Becker, H. (1987). The ethic of care for the self as a practice of freedom. an interview with Michel Foucault on January 20, 1984. *Philosophy & Social Criticism, 12*, 112–131.

Foucault, M. (1966). Les Mots et les choses. Une archéologie des sciences humaines. Paris: Gallimard. English edition: Foucault, M. (2005). *The order of things. An archeology of the human sciences*. London: Routledge.

Foucault, M. (1969). L'Archeologie du savoir. Paris: Gallimard. English edition: Foucault, M. (2004). *The archaeology of knowledge. And the discourse on language* (A. M. Sheridan, Trans.). London: Smith & Routledge.

Foucault, M. (1971). Nietzsche, la généalogie, l'histoire. In Hommage à Jean Hyppolite, ed. Suzanne Bachelard et al., 145–172. Paris: Presses Universitaires de France. English Edition: Foucault, M. (1984). *Nietzsche, genealogy, history*. In P. Rabinow (Ed.), *The Foucault reader*, (pp.76–100). New York: Pantheon Books.

Foucault, M. (1977). Intervista a Michel Foucault. In Foucault, M. *Microfisica del potere, interventi politici*, (pp. 3–28), ed. Alessandro Fontana e Pasquale Pasquino. Torino: Einaudi.

Foucault, M. (1982). Why study power: the question of the subject. In H. L. Dreyfus, P. Rabinow (Eds.), *Michel Foucault. Beyond structuralism and hermeneutics*, (pp. 208–226). Chicago: University of Chicago Press.

Foucault, M. (1984). Histoire de la sexualité (Vol. 2). L'usage des plaisirs. Paris: Gallimard. English edition: Foucault, M. (1990). *History of sexuality* (Vol. 2). *The use of pleasure* (R. Hurley, Trans.). New York: Vintage Books.

Foucault, M. (1984^2). Histoire de la sexualité (Vol. 3). Le souci de soi. Paris: Gallimard. English edition: *History of sexuality* (Vol. 3). *The care of the Self* (R. Hurley, Trans.). New York: Vintage Books.

Foucault, M. (1984^3). What is enlightenment? In P. Rabinow (Ed.), *The Foucault reader*, (pp. 32–50). New York: Pantheon Books.

Foucault, M. (1988). In L. H. Martin, H. Gutman, P. H. Hutton (Eds.), *Technologies of the Self. A seminar with Michel Foucault*. Amherst: University of Massachussets Press.

[11]For Foucault, "the political, ethical, social, philosophical problem of our days" is "to promote new forms of subjectivity through the refusal of this kind of individuality which has been imposed on us for several centuries" (Foucault 1982: 216).

Foucault, M. (2001). L'Herméneutique du sujet. Cours au Collège de France 1981–1982. Paris: Gallimard. English edition: Foucault, M. (2005). *The hermeneutics of the subject. Lectures at the Collège de France* (1981–1982) (G. Burchell, Trans.). New York: Palgrave Macmillan.
Veyne, P. (2008). Michel Foucault: sa pensée, sa personne. Paris: Albin Michel. English edition: Veyne, P. (2010). Foucault. *His thought, his character* (J. Lloyd Trans.). Cambridge: Polity Press.

Commentary: Talking About Time and Whether We Should Measure It

M. Kaufmann

Abstract The emphasis of this short synopsis lies on the way how different authors are struggling with the challenges caused by scientific or frankly physicalistic approaches to the phenomenon of time. It shows how they are trying to maintain something like uncontrollable, immediate human experience of things that are happening in time, something that may get lost if time is completely submitted to precise measurement. Three of the papers discussed are dealing with our way to look at history, while the others have their focus on different ways of describing the gap between individual experience and objective "clock time".

We are confronted with the phenomenon of time—at least if somebody asks us and if we start to think about it as Augustine once did—in a lot of different ways. We may, for the sake of our discussion, posit a first difference between theories dealing with the way how we are looking back on the things that happened to a certain group of people in the past, i.e. with history, and a way how we might describe the different ways in which human beings have to experience time and to deal with it.

Three of the texts presented in this small but interesting collection that I will briefly comment on the following pages are dealing with an adequate mode to unlock the phenomenon of history with our different methods and facing the way how we experience it as human beings. The others have their focus in the awareness of a continuous tension between the concrete, inner, personal, sometimes called phenomenological experience of time, of duration in the terminology coined by Bergson, and an objective, public representation of time, sometimes called clock time which shows a tendency of claiming absoluteness. The relevant authors are differing on the one hand in the way how this opposition is perceived and presented, even if there is a certain, but not unanimous tendency to save individual experience against objectivity and absoluteness.

M. Kaufmann (✉)
Martin-Luther-Universität Halle-Wittenberg, Seminar Für Philosophie,
Emil-Abderhaldenstr. 26/27, 1, 06108 Halle, Germany
e-mail: matthias.kaufmann@phil.uni-halle.de

F. Santoianni (ed.), *The Concept of Time in Early Twentieth-Century Philosophy*,
Studies in Applied Philosophy, Epistemology and Rational Ethics 24,
DOI 10.1007/978-3-319-24895-0_21

Fabrizio Lomonaco shows how Ernst Cassirer takes up the task to deal with history in a way that neither gets lost in a senseless collection of facts concerning a relevant period or institution nor supposes the existence and development of an idea in the Hegelian sense of the word. Instead of an absolute idealism he postulates a critical idealism trying to find the central idea to grasp the "centre of gravity" of a certain époque, in a certain way re-establishing a Copernican turn this time to the approach to history. This way applying his culturalist interpretation of neokantian philosophy to the phenomenon of history, making use of the Kulturwissenschaft, he succeeds or at least tries to evade «both the Scylla of naturalism and the Charybdis of metaphysics». We might say that Cassirer pushes the Vichian project of dealing with history—which conveys the task of approaching the "eternal ideal history" to metaphysics while it is up to philology, numismatics and all the different historical methods to collect historical facts—a step further in direction of nominalism.

The ethical element of history, connected to the insight in the importance of a responsible subject who creates its world of signs and reflections is one of the characteristics of Paul Ricoeur's vision of history, as *Giuseppe Cacciatore* shows. What Ricoeur tries to maintain, to keep alive is the sense of the aporia of time which is closely and necessarily connected to the poetry of narration. That time is indefinite and mysterious does not mean that it is unthinkable or even linguistically unexpressible. It just requires a certain mode of experience and presentation taking distance from any attempt to give an absolute, unifying form of historic knowledge which pretends to be collective subject of history, the acceptance of a pluralist theory of history which is conscious of the fact that historic scripture, writing history, is at the same remedy and poison which may be open to elements of poetry but in fixing things destroys the mystery of time.

Agostino Cera in his analysis of Foucault's "Historic Heterogenies" also refers to two different views on time, one he calls evenemental and the other epistemic. Both are connected or one might say imminent in his way to look at history so fundamental for his "archeological" method and for his rejection of any suprahistoric absolute entity. Evenemental time stems from our first confrontation with singular isolated events, it is without any beginning, without any direction or end. It is the real time of history but simultaneously it marks the limits of archaeological–genealogical research. Epistemic time is completely different, allowing historic consciousness, still free of any sense of history, but opens towards research on provenience and genealogy of historic phenomena. The difference between the two concepts of time and their sometimes aporetic confrontations is closely connected to Foucault's concept of the subject, which is bound to death if it is understood as a cogito or an I but has a renaissance as an ethical being.

Simona Venezia shows in her densely argued contribution on Heidegger's *Being and Time* how his well-known opposition to the traditional view on time in western metaphysics—and physics—as something to be measured objectively remains despite Heidegger's denial connected to Bergsons concept of duration, but finds its anchor point outside of western metaphysics, in the early Christian concept of kairós. "Only kairology can overcome chronology", since only from there we find adequate access to the pathos of existence. Therefore, he discards the traditional subject of metaphysics in favour of the Dasein whose being-in-the world, its relation to the world is always characterized as "interpreting comprehension" (auslegendes Verstehen). This comprehension is closely linked to the idea of projecting the Dasein's future according to its very own potentiality. So, before it gets acquainted with any kind of perception and explanation of things in a theoretical statement it has to understand its world via its interpretation. Because this interpretation happens in the mode of projection it is of "essential temporality", therefore "Only in this way can the *understanding of time* be finally thought as what it really is: the *time of understanding*".

Giorgio Rizzo gives a thoroughgoing analysis of Wittgenstein's reflections on the differences between a phenomenological view on time as something immediately referring to our experiences concerning duration, present, past and future and time as measured by public chronology—using a pair of concepts given by George Edward Moore he speaks of "information time" and "memory time". This way he shows how the philosopher from Vienna identifies the grammatical misconceptions coming up when we are "tyrannized by a metaphor and we are not able to move outside of it" with respect to the usage of the word "time". This may lead us to erroneous identifications of time with its measurement, neglecting the variety of ways in which we refer to the phenomenon of time in our language. It might, however, be exaggerated if the author claims that Wittgenstein "clearly rejects phenomenological language as absurd", that he sees the physicalistic view on time as the only legitimate one and at the same time burdens "the concept of time with a psychologistic emphasis". In his interpretation, Rizzo is relying mainly on the *Philosophical Remarks*, a work of transition between the *Tractatus* and *Philosophical Investigations* in which nevertheless phenomenology—even if not in the strict Husserlian sense is a well accepted part of discourse and reflection. A strong physicalism in my view does not fit quite well into Wittgenstein's type of pragmatism.

Jacques Lacan—as *Marco Castagna* shows—tries to show how logic itself may be temporally structured. He uses for this the sophism of the three prisoners who have to decide whether on their backs there are black or white discs according to which they are bound to be free or held in prison. Lacan shows how the prisoners at least in several cases draw their conclusions in reaction, i.e. temporally after observing the behaviour of the others. One might say that we do not have to interpret this necessarily as a proof for temporal dimension within logic but rather the sophism seems to ask for something similar to case by case analysis in combinatorics. In the end, Lacan himself mentions the difference between *logical time* and *chronological time*. But he still makes another point in differing three kinds of subjects according to different kinds of experience, i.e. an impersonal subject related to initial evidence, a reciprocal subject with a "capability of taking into account the other's reasoning", and an assertive subject which includes the (logical) genesis of the I. This logical mode of self-identification happens in the supposed steps—but is different from phenomenologically experienced duration. To overcome this gap between logical and experienced time Slavoj Žižek proposes a distinction between subjectivity and subjectification. Lacan himself sees the time of subjectivity as the time of language "the anticipation of the ending of the sentence". In my view, there is the remaining problem that language is always tied to temporal linearity, whereas our subconsciousness proceeds in simultaneous activities. Maybe this was one of the reasons why Lacan turned to topology in his later works.

Part III
Science and Logic of Time

Introduction

Mirko Di Bernardo

Bergson, echoing Augustine, sustained that we "do not think of real time", but rather live it inasmuch as "life transcends the intelligence". In effect, there has been no significant thinker, from antiquity until our time, who has not made the treatment of the mystery of time an essential moment of their philosophizing. Take, for example, Parmenides and Zeno, who consider time "subjective semblance", or Kant and German idealism who transfigure it into the "pure form of sensibility" or "intuition"; think of Democritus and Epicurious who degrade it to the level of an "accident of accidents", until reaching the conventionalism of Mach where it becomes a "useless metaphysical concept"; take Plato, who transposes time as the "mobile image of eternity" and Aristotle who posits it as the "number of movement according to before and after"; or still yet, consider Leibniz who considers space and time as conceptual apparatuses that describe the interrelations between events, and Plotinus who observes how time is not the measure of movement of the universe, but rather, the movement of the universe is in time. Finally, think of Newton's absolute time, according to which time "flows uniformly" in itself and by nature "without relation to anything external". Such definitions of time, as is well known, really only constitute images, translations, or metaphors since they all presuppose, upstream, the very idea of time. A new and operational (or operative) definition of the concept of time can be found, however, in the twentieth century, in Einstein's theory of relativity, which strips the notion of any metaphysical content, and ends up transforming time from the cause of the clock's movement to the effect of the same. The same concepts of past, present, and future become purely operational, contributing, moreover, to placing the very notion of causality into definitive crisis. One can thus see how all this has contributed in a decisive manner to the launch of a real gnoseological revolution, where every ontological foundation, one after the other, is destined to a slow but inexorable breakdown. Although with the publication of Euler's *Reflections on Space and Time*, the "nominalist" position—which many critics trace back, as its most ancient point of reference, to

M. Di Bernardo (✉)
Department of Business Government Philosophy Studies, University of Rome
"Tor Vergata", via Columbia 1, 00199 Rome, Italy
e-mail: diber.mirko@gmail.com

F. Santoianni (ed.), *The Concept of Time in Early Twentieth-Century Philosophy*,
Studies in Applied Philosophy, Epistemology and Rational Ethics 24,
DOI 10.1007/978-3-319-24895-0_22

Parmenides and the Eleatic school—would seem buried forever in virtue of the apparent victory of the "realist" position—which can be traced back to Heraclitus and Plotinus—yet two centuries away from the Netwonian recovery, the dawn of relativity not only freshly places in doubt the ontological reality of space and time, so long assailed over the centuries, but even goes so far as to consider the very division between past, present, and future a mere illusion.

In contrast with classical Newtonian physics, Einsteinian theory does not place matter within an indifferent reticulated space, but describes a reciprocal relation between matter and the metric properties of space-time, which absorbs it. Einstein's restricted theory of relativity remains linked to a metrics characterized by a curvature nothing like that of Euclidean space; however, this metric does not space alone, but also time. Restricted relativity defines a new invariant, a new distance no longer between two points, as in Galilean and Newtonian physics, but between two spatiotemporal events. Different observers moving toward each other in uniform, rectilinear motion can no longer agree either on the distance between two events or on the time that has elapsed between them; they can only agree on the spatiotemporal interval that separates the events. It is this physical quantity, or new invariant, that is maintained when one passes from one inert observer to the other. This implies that each observer will see a source of light shift in the void in such a way that the interval mentioned above is canceled out. For all these observers, therefore, the speed of light will have the same value. It is in relation to this four-dimensional continuum that Einstein reinterprets the acceleration caused by the forces of gravitational interaction. From the definition of the curvature in one region of the universe, there is derived the definition of the motion of a test body in this region. Einstein's equations therefore describe a space-time that reacts in the presence of matter, a matter sensitive to the curvature of space-time. More precisely, the fundamental equation of general relativity links two mathematical objects called tensors: the metric tensor which describes the curvature of space-time in the region of the universe in question and the impulse-energy tensor that defines the content of matter of that region in terms of density and pressure. General relativity, which is at the basis of the standard cosmological model, has introduced the revolutionary conception of a relationship between space-time and matter; however, it has conceived of this relationship essentially as symmetrical, where the presence of matter causes a curvature of space-time and this, in turn, causes the movement of such matter. Not unlike the Netwonian theory of motion, of which it is the heir, general relativity gives no meaning to irreversibility and in particular does not allow us to explain the considerable production of entropy which, as we know today, seems to have marked the birth of the current universe. While Einstein dreamt of a geometry capable of unifying the physical laws and of tracing back the set of physical–chemical processes to this basic geometrical truth, allowing us to conceive of the indissoluble unity of space-time and matter, today we know, thanks to the important theoretical contributions offered by Prigogine and others scholars at the level of the science of dissipative processes that produce entropy (the physics of dissipative systems) and of the science of complexity, that such a unity does not mean equivalence. In the perspective offered by the Russian scholar since the late

1970s, the creation of entropy-bearing matter cannot be traced back to a reversible phenomenon any more than the excited quantum atom can be traced back to mechanics. In this perspective, it is the very production of entropy that constitutes the real "price" of the passage to the existence of our universe and that constitutes, therefore, the difference between this material universe and an empty one. The possibility of defining such a difference and the passage to existence has recently led to some attempts to generalize Einstein's equations, allowing us to describe an irreversible process of the creation of matter. Thus, according to the Brussels school, one can substitute for the initial singularity sustained by standard model, an instability that leads to a simultaneous creation of the matter and entropy of the universe. If Einstein's dream conceived becoming as an obstacle for physics and the arrow of time (symbol of the one-directional nature of time that found no "abode" in the physics of the early twentieth century) as an illusion to be overcome, today becoming would seem to irrupt precisely where this dream had found its most evident expression, that is to say, in the symmetry established by general relativity between matter and space-time. The initial instability, in agreement with Prigogine, makes of the universe the product of a breakage of symmetry between space-time, on the one hand, and matter, on the other, placing moreover the birth of the material universe under the sign of a radical irreversibility: the laceration of the uniform fabric of space-time from which matter and entropy are simultaneously and constantly generated. Starting from some studies done mainly at the level of the thermodynamics of nonequilibrium and in virtue of some very recent experiments of particle physics that would show the existence of the arrow of time,[1] the radical conceptual transformation concerning temporal irreversibility has gradually penetrated into almost every level of physics, today allowing us to glimpse the possibility of a new coherence, articulated around that becoming that the physics of yesterday interpreted as an obstacle and which instead Bergson considered to be that "whole of the same nature of the Self" that can be grasped through an "ever more complete deepening in oneself".

To be sure, the natural sciences have been lacerated from the very start by the opposition between the irreversible time of phenomenological descriptions and the intelligible eternity of the laws that permit us to interpret these phenomenological descriptions. Today metamorphosis (becoming) and invariance (intelligibility) are not so clearly contraposed, even if the problem of eternity has not disappeared from the international epistemological panorama. Deriving from the alliance between theoretical invention and experience, there is experienced within contemporary science an intense and constant generative opposition of these two complementary

[1]The observation was made by two teams of scientists, from the CERN of Geneva (CPLEAR experiments) and the Fermilab of Chicago (KTeV experiements), respectively. Such observations, which became significant only at the end of the 1990s, regard processes that occur in the sector of K mesons (Kaons) and their antiparticles (anti-Kaons) and would seem to demonstrate for the first time that matter "distinguishes" (at the microscopic level) between past and future inasmuch as there was observed a violation of symmetry (invariance) by temporal inversion. The result contradicts the conviction expressed by Einstein that time at the subatomic level does not exist.

souls represented by freedom and limitation. Its inventive force is manifested in the creation of new languages, in particular formal and symbolic languages, which allow us to introduce distinctions inaccessible to natural language. It is not a question, in other words, of denying the physics of eternity, but of responding to the challenge that its success has constituted. From Galileo to our days, physics has been guided by an ideal of perfection—Leibniz's principle of sufficient reason—whose reverberations have influenced decisively, especially in the last century, all the sciences, including logic, epistemology, and the cognitive sciences. The reversible equality between cause and effect has constituted an almost invisible constraint that has led to Galilean physics, quantum mechanics, and the theory of relativity, giving birth moreover at the epistemological level to the linear approach or to reversibility inspired by more or less refined forms of determinism and reductionism. Just as Kepler renewed the cognitive ideal of astronomy, breaking the circle that had lead from Ptolemy to Copernicus, Prigogine, and others scholars have contributed to fracturing the circle of sufficient reason, creating a new mathematical language capable of making intelligible the irreversible processes and events that traditional physics had limited itself to saving through phenomenological approximations. Hence the possibility to identify the time of complexity as the complexity of time: time once again becomes the undisputed protagonist of the phenomena—and therefore of the scientific analysis—of observable dynamics and of those dynamics not yet subject to a completed measurability. A time that sheds light on the argument of complexity, today the starting point for theoreticians of nonlinearity and for the search for a knowledge that, in the past, proved to often simplified an ignored. As difficult and risky it is to consider complexity theory a theory, it highlights first of all the intrinsic irreversibility of every natural phenomenon, be it molecular, cellular, social, or digital. Complexity theory is the invention of new languages, the opening of new possibilities of thought and expression of the reality that we live. Absolute being seems thus to get lost, to make room for becoming; at this level, however, the arrow of time imposes itself as a new thought of eternity, that arrow of time that had been judged relative only to the approximate character of our knowledge and which, once again, we now find instead as the unconditioned condition of all the objects of physics, from the hydrogen atom to the universe, not only enabling us to conceive of the processes that share the same future, but also giving us the possibility to glimpse at the very roots of our universe the intrinsic difference between past and future, without which it would prove impossible to think, speak, and act. The definition of the instantaneous state thus breaks the symmetry between past and future a breakage of symmetry that is further amplified in a manner directly proportional to its evolution. Bergson, to express in realistic terms the solidarity that unites us to the time of things, which is translated by the laws of probability, had written that "we must wait until the sugar melts". It is this solidarity that the dynamic of chaotic systems explicitly affirms.

Therefore, in light of all this, according to the nonlinear approach proper to the complexity theory envisaged by Prigogine, like Augustine, we do not know what time is, but it proves possible to trace back the laws of motion to integral definition of Aristotle, where the intrinsic measure of these laws imposes the perspective of before and after. Unlike the conception of motion of Galileo and his successors, according to which at each instant the dynamic system is defined by a state that contains the truth of its past and of its future, the Russian scholar's conception of motion offers width to the instant and joins it to becoming, so that every instantaneous state is the memory of a past that allows us to define only a future delimited by an intrinsic temporal horizon. This weft presents itself, at the same time, as creation and revelation. As the continuous creation of new forms of autonomy and, contemporaneously, as the continuous revelation of new levels of generative power: an emergence of continuous novelties able to shape consecutively and in a close-together manner the determinations (or schemas) of time, which form, in turn, on the basis of precise mathematical modules, the varied and bound expression of the language of life.

The chapter presented in this section retraces, albeit in a very general fashion, the interwoven threads of the two main approaches that animate the contemporary epistemological and scientific debate with particular reference to the concept of time, that is to say, the one of linearity (the circle of sufficient reason common to classical, relativistic, and quantum physics) and the other one of nonlinearity (the physics of dissipative processes and complexity theory), illustrated in these first pages through the articulated comparison between Einstein's theory of relativity and Prigogine's theory of dissipative entropy-producing processes. Some issues crop up again and again as dominant themes in lively contraposition like; take, for example, the issues of instability and the event (temporal irreversibility) in opposition to the circle of sufficient reason (temporal reversibility), which presupposes the possibility of defining the cause and effect, between which a law of evolution would establish a reversible equivalence. Instability resists this ideal, opening a new field of problems in which the event, that is, the historicity of evolutive processes, plays a central role. In all the contexts addressed here not only by physics and epistemology, but also by logic, psychology and linguistics (though under different forms), one finds this dialectical process of generatively juxtaposing complementary souls, between the event (time) which creates a difference between the past and the future and sufficient reason (eternity) which attempts to define them as equivalents. According to the positions sustained by the various authors, in virtue of this complex dynamic regarding the comparison between symmetry and asymmetry, the possibility emerges of overcoming the opposition between the object subjected to the categories of sufficient reason and the subject which, by definition, should elude them. Finally, there are interesting attempts at integrating Parmenides' being, eternally identical to itself (the atemporal laws of physics) with the becoming that alone, as the Stranger in Plato's *Sophist* sustains, allows one to give a sense to life and to the intelligence that learns (the incompressible complexity of the living being).

Rocco Pititto highlights, in his essay, the fundamental contribution offered by Guillaume to European linguistics by considering human language under the viewpoint of the temporality of mental operations, a hypothesis that constitutes the nucleus of his conception, a time that is not immobile, but subject to a dynamism, that of life itself. Guillaume sustains that the human mind experiences time, but not its representation and must therefore avail itself of spatial constructions to represent it. The usual representation of the grammar of a divided linear time leads to segments, past and future, but this is not enough, because it shows us a time that has already been constructed, but not a time in the act of constructing itself in thought, that is to say, operative time understood as an infinitesimal time of mental operations that inhere in the construction of language. In this sense, then, thought becomes the place of the definition of time, while time becomes thought's place of action.

Salvatore Principe, in his essay, accurately expounds the epistemological novelties introduced by Einstein's restricted theory of relativity regarding the concept of time. Underlining how from Einstein's theory one understands that, in nature, isolated systems are only an abstraction or are particular cases, while the rule is that of open systems that exchange energy with their surrounding systems, and are, thanks to this, in constant evolution. Principe allows one to glimpse how the line of thinking regarding complexity and dissipative systems was opened precisely by that initial putting into crisis of classical physics/metaphsyics. That relativizing of the systems of reference and the putting into crisis of the concept of invariance was the beginning of quantum physics and complexity theory.

Michele Malatesta illustrates the relationship between verbal tenses and temporality in Reichenbach's thought. According to the German philosopher, verbal tenses express relations between three temporal parameters: the time of the event, the time of emission, and the time of reference. His most original contribution consists in the attempt to rationalize verbal tenses while disregarding the present tenses of the various languages, which often arise rhapsodically on different or even diverging bases. Although Reichenbach's inquiry gives rise to a vast flourishing of two kinds of research concerning, respectively, tense logic and temporal logic, as Malatesta notes, the refusal to distinguish between the sentence (understood as a linguistic expression) and the proposition (conceived as the meaning of the sentence) impeded Reichenbach from taking the final step, realizing the passage from verbal tenses to the condition that founds them.

Nicola Grana analyzes the concept of time in Prigogine, maintaining as a theoretical horizon of reference the fundamental idea that is the basis for the works of the Brussels school, according to which irreversibility would prove closely linked to the notion of dynamic instability. Grana carries out his analysis by illustrating the cosmological model proposed by Prigogine of a universe that exhibits both an age (an origin) and an arrow of time. In this perspective, the symmetry of the relations that Einsteinian cosmology established between space-time and matter, inherited from the Newtonian theory of masses in gravitational interaction, is broken: matter is distinguished from space-time by its bearing the entropy of the universe. Its existence is no longer a given, as the standard model presupposes, but is rather the product of an irreversible process of creation.

Gianluca Giannini addresses the topic of the destructuring of the realist conception of time. The end of time, the title of his contribution, alludes not only to the end of progressive time, but to the end of temporality conceived as the virtuality of a past that is no more and a future that is not yet. The existential, ontological, and physical condition of time defines our identity as subjects and as a species. Giannini takes his point of departure from Deleuze's consideration in *Time and Synthesis* and from the link with Kant. Deleuze, in particular, warns that time cannot be defined. Every attempt in this direction defines the object through its attributes, but fails to clarify its nature. A real definition of time appears even more difficult when—after the Einsteinian revolution—time seems to have "evaporated" and with it, the possibility of defining and determining human identity.

Note of Introduction

Gianluca Giannini

The end of time? (from Barbour to Hawking... and back) In the *Timaeus* Plato writes:

> Time is a moving image of eternity (37d).

According to this philosophical position, in the Eastern Tradition, along several centuries, after the "revolution" of linear time introduced by Christianity, we tend to believe that destiny is not fixed and that all time past fades into oblivion. But could this speculative movement be only an illusion?

In the last century, Albert Einstein had already demonstrated that temporal reality is relative to each object in the universe, and that time is a "subject" inseparable from space. Even specialists who synchronize time in the world are aware that the world is handled by an arbitrarily stipulated ticking, as clocks are not able to measure time at all. Apparently, the only alternative is to sink into a "temporary illusion" of this infinity, knowing that there is a space where our past still exists and what we do does not change. As Einstein himself would say in one of his last *Letters*: "People like us, who believe in physics, know that the distinction between past, present and future is only a stubbornly persistent illusion".

Therefore, time does not exist?

It is unreal?

Saying that time is unreal could mean denying so-called "temporal becoming": the idea is that past and future things, events and states of affairs (or, however, one conceives the material contents of spacetime) are just as "real" as present ones. On the other hand, saying that time is unreal could mean "the opposite", i.e. presentism. The idea is that only the present is real: past and future things are unreal. As regards the main idea of presentism, it does not matter how one conceives the material contents of spacetime, though of course in more precise versions, it can matter.

G. Giannini (✉)
Philosophy Section, Department of Humanities, University of Naples Federico II, via Porta di Massa 1, 80133 Naples, Italy
e-mail: gianning@unina.it

F. Santoianni (ed.), *The Concept of Time in Early Twentieth-Century Philosophy*, Studies in Applied Philosophy, Epistemology and Rational Ethics 24,
DOI 10.1007/978-3-319-24895-0_23

Thus the debates about these two positions turn on the contrast between the real and the unreal.

But there is, now, a third conceptual form. An interpretative form, capable of dissolving the reasons of the debate between the real and the unreal. We can speak of spontaneity.

What does it mean?

Spontaneity presupposes the idea of a set of many possible courses of history, where each course of history is a "block universe". But Spontaneity then proposes that unbeknownst to us, the actual history jumps between disparate instantaneous states.

The most painful thing to humans, as Eastern philosophies outline, would be to try to break the fixed mould. The wise one, who follows the predetermined course, would be a happy face amid the cosmic chocolate custard who tries to live our unique and extremely tiny "nows".

Science and Logic of Time Maps

Flavia Santoianni

1 Einstein and Prigogine Theories Atlas El Map

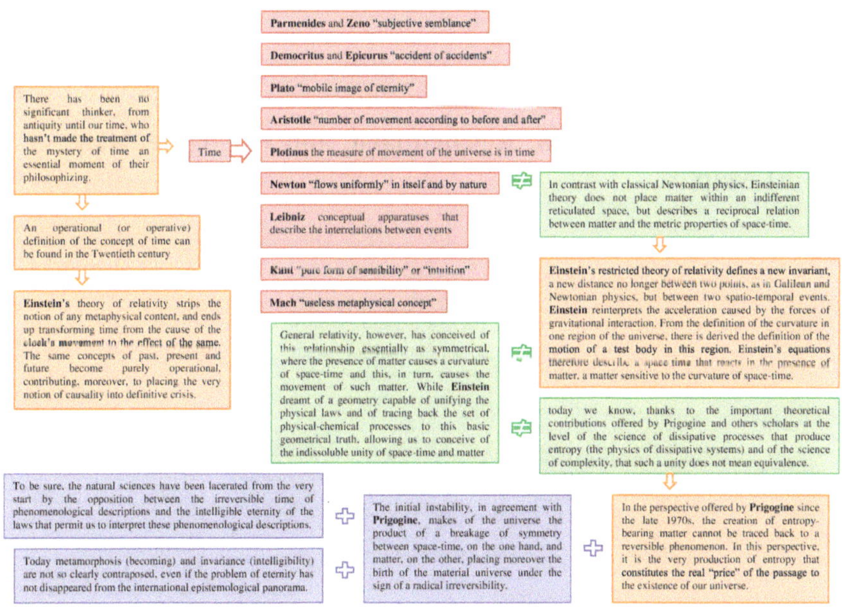

Time in Einstein and Prigogine Theories – EL map – Flavia Santoianni

---

F. Santoianni (✉)
Dipartimento di Studi Umanistici, University of Naples Federico II, Naples, Italy
e-mail: bes@unina.it

F. Santoianni (ed.), *The Concept of Time in Early Twentieth-Century Philosophy*,
Studies in Applied Philosophy, Epistemology and Rational Ethics 24,
DOI 10.1007/978-3-319-24895-0_24

2 Time in Contemporary Science Atlas El Map

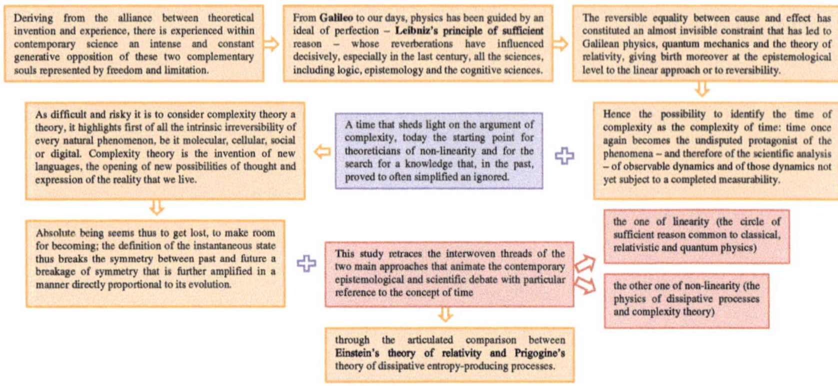

Time in Contemporary Science – EL map – Flavia Santoianni

3 Guillaume, Einstein, Reichenbach, and Prigogine Atlas El Map

Time in Guillaume, Einstein, Reichenbach, and Prigogine – EL map – Flavia Santoianni

4 Barbour Atlas El Map

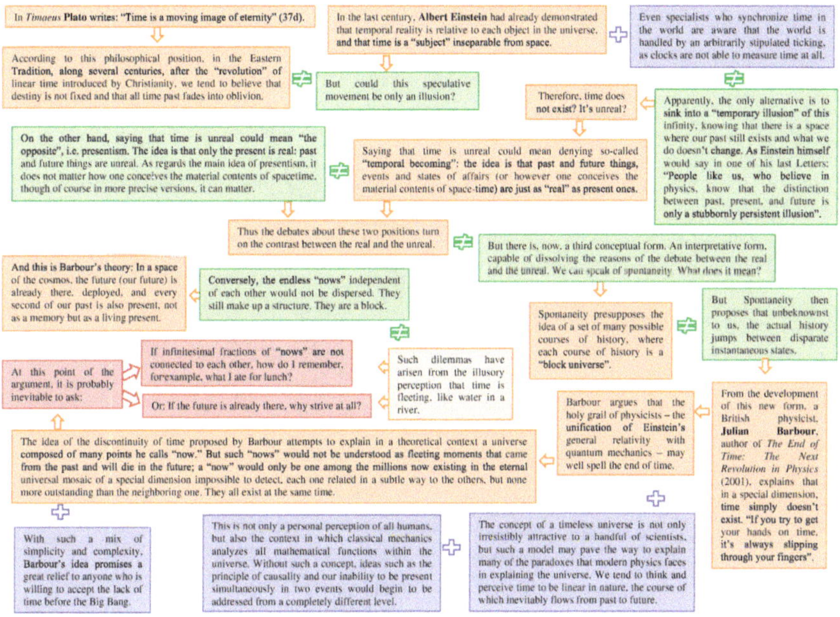

Time in Barbour – EL map – Flavia Santoianni

5 Atlas Map

Atlas map – *Philosophers are located near their birthplaces. The Geographical Boundaries of Countries May Differ in Comparison with the European Geography of the Early Twentieth-Century*

The Linguistics of the 1900s from Ferdinand de Saussure to Gustave Guillaume Between Synchrony and Diachrony

Rocco Pititto

Abstract According to Gustave Guillaume, a linguist endowed with incontestable speculative depth, though misunderstood by the linguists and philosophers of his time and rather ignored in linguistic textbooks, language has a temporal architecture, determined by the articulation of time, which from the present, is projected into the future, while having and maintaining its roots in the past. The present is only the interval between the past and the future. As such, time, however, cannot be represented by way of itself: it requires a representation that can only be made via spatial instruments. Guillaume has the merit of having considered human language in the temporal dimension of thought operations, causing it to become an interpretive paradigm of language itself. This is a paradigm able to give a sufficiently well-founded explanation for a series of linguistic phenomena, otherwise lacking sufficient explanation. The theoretical principles on which Guillaume founds his discourse on comprehension, refer to three aspects in particular: operative time, the central concept of Guillaume's approach to the problems of language, the reciprocity of the relationship between language and discourse, and the idea of succession in the process of constructing language.

The European linguistic and philosophical culture has a debt in respect to Gustave Guillaume (1883–1960), a gifted linguist with considerable speculative depth, original, and innovative, but misunderstood by the linguists as well as by the philosophers of his time and ignored in the manuals of linguistics except for a few exceptions (Moignet 1981; Soutet 1998).

Gustave Guillaume remains an "only slightly more than unknown author" in Italy (Martone 2006), as well as in France and elsewhere in Europe. His work has long been ignored, and the attention of scholars, since the 1950s, has focused mainly on the general linguistics of Ferdinand de Saussure, from which, for that matter, Guillaume had taken inspiration, already in the late 1920s (Wilmet 1978; Hirtle 2007). Rather than conflict with the language of De Saussure, Guillaume

R. Pititto (✉)
Philosophy Section, Department of Humanities, University of Naples Federico II, via Porta di Massa 1, 80133 Naples, Italy
e-mail: pititto@unina.it

F. Santoianni (ed.), *The Concept of Time in Early Twentieth-Century Philosophy*, Studies in Applied Philosophy, Epistemology and Rational Ethics 24,
DOI 10.1007/978-3-319-24895-0_25

completes it, following in the study of language an inquiry procedure between the synchronic and the diachronic, and by linking the description of the language and its grammatical components with the identification of its related mental correlation with the study of the occurrence of the meaning. The description of language and the study of meaning refer to two different types of knowledge, to linguistics on the one hand and to philosophy on the other, in particular, the philosophy of mind. There is on the part of Guillaume the clear statement about the need for a continuity of interests and perspectives between the planes of the two fields of knowledge, a circumstance which would have led to further developments in the field of linguistics, if only he had proceeded to operate a kind of bond between two different but not irreconcilable positions. An integration of the two aspects of research on language would be possible with positive results for a broadening of the research in a more philosophical context. It was an opportunity of methodology and of perspective that was not at all extraneous to him. The qualification of a linguist that has been attributed to him appears to be too reductive to give full account of the complexity of an open research that finds in the faculties of the mind the first reference point of his conception that is linguistic and philosophical at the same time. Linguistics is the science

> qui introduit le plus avant à la connaissance des moyens avec lesquels notre pensée parvient en elle-même à la saisie claire de ses propres démarches (Guillaume 1973, p. 31).

With reference to Ferdinand De Saussure, one cannot ignore, on the other hand, how the linguist from Geneva, along with his school, was long considered the leader of that linguistic research subsequent to the debate that took place in the nineteenth century, and even before in the second half of the eighteenth century. With his views, he would have caused, especially in the second half of the twentieth century, a kind of unification of knowledge, operating a rapprochement between linguistics and the most diverse humanities and providing the latter with a method of investigation borrowed from his own linguistics. Following the lesson of De Saussure, linguistics itself became an indispensable method to describe and characterize the most diverse fields of knowledge. The paradigms of Saussurian linguistic concepts could also be referred to the most diverse fields of knowledge, from anthropology to architecture, from philosophy to social sciences, from literature to psychoanalysis, from morality to meaningful practices. The birth of semiology, first, and semiotics, later, represented the triumph of the linguistics of De Saussure, because it made possible the transformation of linguistics itself into a general conception of culture, as if it were a new metaphysics in the context of the society of the twentieth century, which had even rejected the return of all forms of metaphysics. It was only an illusion, which lasted a little longer, because even the linguistics of De Saussure has known and is knowing today its decline, lost in the many structuralisms, each of which advances the claim of being the most faithful interpreter of the lesson of the master (Pititto 2012). The crisis of the post-Saussurean linguistics of the twentieth century is at the origin of today's new interest in the linguistic conceptions of Guillaume.

Certainly, a greater attention on the part of scholars for the ideas of Guillaume could have, perhaps, resulted in the narration of a different story in the context of the development of the linguistic sciences of the twentieth century, going beyond the doldrums of a linguistics that was too "unbalanced" in favor of the Saussurian *langue-parole dichotomy* (Albano Leoni 2009, p. 17). Guillaume rewrites the Saussurean *langue-parole* equation by replacing the *parole* with the term *discours*, so that the act of language is formed as a unit of *langue-discours*. In the act of language, Guillaume identifies the level of power (the *langue*) and the level of the act (the *discours*). Both levels are structured under the sign of temporality, understood by Guillaume as an interpretative paradigm of the linguistic act (Soutet 1998, p. 141).

1 Gustave Guillaume and Temporality in the Language: An Interpretative Paradigm

The different systematization in all languages of the articulation of time in the present, past and future is explained by linguists as the result, more or less accidental, of the mechanical development of language. According to Gustave Guillaume, this explanation, if it appears valid from a historical point of view, is not so from a theoretical point of view.

> Comment concevoir, en effet

he asks

> que du seul développement mécanique du langage puisse résulter un édifice aussi abstraitement systématique que celui du temps dans l'universalité des langues?.

This statement, firmly rooted in the systematization in the temporal sense of languages, on which many linguists agree, manifests an inconsistency,

> à laquelle on n'échappe que si l'on suppose le système du temps capable de s'accommoder, par le jeu de transformations intérieures n'en altérant pas l'unité d'agencement, aux conséquences matérielles du développement mécanique du langage (Guillaume 1965, pp. 1–2).

The differentiation between an action already accomplished and an action yet to be done or an action in the course of being performed requires on the part of the speaker a remarkable capacity for abstraction and cannot be explained as a purely accidental fact, which affected the development of the language.

Guillaume incorporates the Saussurean *langue/parole* distinction, giving it, however, a highly modified version in the form of *langue/discours* and taking language as a unitary result of the same *langue/discours* relationship (Soutet 1998, p. 141; Stecher von 2012). The change is not only terminological, because it refers to a semantic field, which in part can be compared to that of Saussure. It is temporality which determines this change of meaning by assigning to language and to speech a connotation of the temporal kind. According to Guillaume, human language presents itself from the beginning with two vocations, on the one hand that of making itself *language* and on the other that of making itself *speech.*

> Il se fait langue quand, infléchi en direction de la puissance, il s'institue et, du même coup, se délie de la condition de moment, et discours, quand infléchi en direction de l'effet, il se présente non institué, lié, assujetti à la condition de moment (Guillaume 1973, p. 217).

And, again

> On part d'une condition de langue unique habile à porter, en discours, des conséquences diverses, très variées. Discours et langue sont des termes corrélatifs. Pas de discours sans langue. Au cas où je ne posséderais pas la langue en moi, pré-construite, j'en serais réduit aux possibilités quasi nulles du langage improvisé, ayant à inventer des moyens d'expression dans le moment du besoin. Je serais incapable de produire un discours lié, suivi (Guillaume 1971, *Leçon* 2 Dec. 1948).

The making itself language and the making itself speech imply different modes in the articulation of language, characterized by a suspended movement, using a phenomenological terminology, amid the filling of meaning and the making of meaning. The two moments of the process of the change of language take on no less importance in terms of the representation itself.

As Guillaume does not fail to underline, language and speech play different roles in terms of representation. If language is already in itself a representation system, speech is its realization.

> La langue est un système de représentations. Le discours un emploi, aux fins d'expression, du système de représentations qu'est en soi la langue. [...] La représentation du temps est une construction architecturale que la pensée édifie au plus profond d'elle-même, *n'ayant d'autre objectif que de la réussir,* d'en faire un ouvrage cohérent, en correspondance avec l'expérience que l'esprit humain a du temps à une époque de civilisation donnée. [...] Mais si changée que soit, d'un idiome à l'autre, la représentation architecturale du temps au fond de la pensée, le discours, dont les fins ont quelque chose de constant, en obtient une expression relative à des oppositions dont un caractère frappant est de se répéter universellement, sous la seule réserve que l'état de civilisation ne soit pas, fondamentalement, trop différent (Guillaume 1964, pp. 208–209).

Language as a unit of *langue* and of *discours* is not the sum of distinct faculties, but a process that is divided into two distinct phases of *langue* and *discours,* and is characterized as a continuous passage from the plane of *langue* to that of *discours*. This step, however, would not be conceivable without a temporal support. In the linguistic phenomenon, a task is accomplished that is actualized in an infinitesimal duration. It is an infinitesimal temporality that Guillaume considered effective and to which he gave the name of *operative time*, the very time of thought in the act of language. Language has, therefore, its temporal architecture, and this is determined by the articulation of time, that from the present projects itself into the future, having and maintaining its roots in the past. The present is only the interval between past and future. Time, however, cannot be represented from itself: it needs a representation, which it can only take up from spatial instruments. Thus, representations of time are but a *spatialization of time*. The representation of time, called by Guillaume *chronogénèse* and *chronothèse*

> est, au vrai, une *spatialisation du temps,* […] le temps, non représentable à partir de lui-même, emprunté ses moyens de représentation à l'espace, et qu'il est, lui, recouvert d'une représentation spatiale, en l'absence de laquelle nous ne le connaîtrions que comme expérience: ce qui ne serait pas le connaître (Guillaume 1973, pp. 22–23).

Gustave Guillaume's merit was to have considered human language in the temporal dimension of the operations of thought, making it an interpretative paradigm of language itself. It is a paradigm able to give a sufficiently well-founded explanation to a number of linguistic phenomena, without which they would not find a sufficient explanation. The theoretical principles, on which Guillaume founded his discourse on comprehension, refer to three aspects in particular: the operative time, the central concept of Guillaume's approach to the problems of language, the reciprocity of the relationship between language and discourse, and the idea of succession in the process of the construction of language. If the operative time is the infinitesimal time of mental operations that contribute to the construction of language, the reciprocity of the relationship between language and speech allows one to highlight the moments of the construction of speech along with the linguistic changes in the transition from speech to language. Finally, the process of the construction of language passes through its semantic construction (the *ideogenesis* or *semantogenesis*) as well as the establishment of morphosyntactic operations of the noun phrase and verb phrase (*morphogenesis*) (Begioni 2010, p. 125). The phenomenon of language is considered in its diachronic aspect, rather than in the synchronic, in its becoming word in everyday use that the speakers make of it.

From the identification of these characteristic features of language comes a linguistic model in which and with which Guillaume founded and summarizes his conception. It goes by the name of "psychomecanics of language" or "psychosistematics of language," and its characteristic is to be traveled by the thought of the time, as that which binds it to the dynamism of life and its many changes (Lowe 2007).[1]

> La psycho-systématique n'étudie par les rapports de la langue et de la pensée, mais les mécanismes définis et construits que possède la pensée pour opérer une saisie d'elle-même, mécanismes dont la langue offre une reproduction fidèle.[…]. Or, ces moyens qu'a la pensée de se saisir elle même dans sa propre activité – quelle qu'elle soit – présentent, on aura l'occasion de le constater, un caractère mécanique. Ce en présence de quoi on se trouve, ce sont des psycho-mécanismes dont le principe constructif est la recherche d'une commodité de saisie et la recherche aussi, dans le système, d'une saisie instituée, d'une économie supérieure procurant cette commodité (Guillaume 1973, pp. 94–95).

In the linguistic conception of Guillaume the dimension of time takes on a decisive role.

[1]Guillaume distinguishes between "thinking properly said" and "the power that it has to perceive itself". The two cannot be confused.

> Il importe en toute question linguistique ayant trait au temps de faire le départ entre le mouvement du temps dans la pensée et le mouvement de la pensée dans le temps. La pensée est le lieu de définition du temps, mais le temps est le lieu d'action de la pensée (Guillaume 1964, p. 60 note 8).

Guillaume's model, by the very fact of establishing a correlation between linguistic facts and mental phenomena, takes on the value of a hermeneutic cipher significant enough of the linguistic phenomena in their complexity to be both linguistic facts and mental facts. As such, it can be used by scholars as a useful tool for understanding the phenomenon of language and its transformations underway and, again, for the prediction of the possible transformations that the language itself can have in the future. The great merit of Guillaume is to have considered human language under the perspective of the temporality of the operations of thought, a hypothesis which forms the core of his conception, a time that is not immobile, but subject to a dynamism, that of life itself (Begioni 2010, pp. 125–136). The theoretical approach of the concept is centered on the concept of operative time, an infinitesimal time of mental operations relative to the construction of language. From the concept of operating time Guillaume brings forth the relationship between language and discourse, a relationship that links the construction of speech with linguistic structural changes from the speech towards language (Begioni 2010, p. 125).

Being ahead of its time to be taken into consideration, the psychomechanics of language can be a possible response now, when cognitivism seems to have already exhausted all its capabilities and, in the same time, a passage has occurred, not without meaning, from the philosophy of language to the philosophy of mind. The presupposition of this passage is given by the fact that a purely linguistic consideration of language does not explain that mental aspect that constitutes it in its fundamental characteristics. On the other hand, it is already settled how the merely descriptive activity of the linguistic phenomenon is not sufficient, if one wants to achieve a more complete understanding of it. According to Guillaume, the moments of each linguistics are the "observation" and "reflection" of the inner universe. It is not enough to observe the simple linguistic act, already accomplished; it is important to reflect on the process through which the linguistic act relates to the operations of thought (Guillaume 1973, p. 37). The temporal aspect is one with the idea of the process.

The idea of the process is fundamental in the articulation of the views of the French linguist. The idea of the process is linked to the other idea of the interception of thought. If psychosystematics studies the mechanisms that thought possesses and puts things in place to operate an "interception" of itself, mechanisms for which language provides a faithful reproduction, psychomechanics corresponds to the search for a comfortable capacity of "interception" (Guillaume 1973, p. 223).[2]

---

[2]As Roch Valin explained in *Principes de linguistique théorique*, the term "interception" translates into the term "saisie". In French "saisie" has a double meaning: it can be understood as the meeting point or point of interception of two different levels, and, also, as the act of grasping the result of this interception.

> Ce qui se conçoit, une toute première nécessité de l'acte d'expression étant que la pensée ait acquis la puissance de se saisir elle-même. Sans saisie de la pensée par elle-même, pas d'expression possible (Guillaume 1973, pp. 94–95).

Only if the thought is able to reflect on itself there is an act of expression in the language of the individual. The linguistic act has its origin not in the mouth, but in the brain of an individual who reflects on himself and transforms into language the representations of his experience. The referent of language is always mental, and as such is always elusive.

> La langue se compose de résultats sous lesquels il s'agit de découvrir, afin de rendre raison des choses, l'opération de pensée créatrice (Guillaume 1973, p. 223).

The golden rule that Guillaume forces himself to follow is to transform a found result, such as a noun or an adjective, into a process, a genetic process. The process of substantivization as well as of adjectivization, as a fundamental mechanism of the language, is the object of the analysis, not the noun, or adjective.

> Autrement dit la règle d'or [...] c'est la réversion du résultat constaté en procès – en procès génétique. C'est ainsi qu'au substantif qui est dans la langue une chose visible, un résultat, on a opposé le procès, nécessaire et antécédent, de la substantivation; et à l'adjectif, lui aussi visible et lui aussi tenu pour un résultat, le procès d'adjectivation (Guillaume 1973, p. 223).

References

Albano Leoni, F. (2009). *Dei suoni e dei sensi, il volto fonico delle parole*. Bologna: Il Mulino.

Begioni, L. (2010). Psicomeccanica del linguaggio e temporalità. In M. Castagna & S. De Carlo (Eds.), *Lo spazio della parola* (pp. 125–136). Napoli: EDI.

Guillaume, G. (1973). *Principes de linguistique théorique*. Québec City: Les Presses de l'Université Laval et Paris Klincksieck.

Guillaume, G. (1965). *Temps et verbe. Théorie des aspects et des temps*. Paris: Champion.

Guillaume, G. (1971). *Leçons de linguistique 1948–49*. Québec City: Les Presses de l'Université Laval et Paris Klincksiek.

Guillaume, G. (1964). *Langage et science du langage*. Québec City: Presses de l'Université Laval et Paris Librairie Nizet.

Hirtle, W. (2007). *Language in the Mind. An Introduction to Guillaume's Theory*. Montreal, Kingston: McGill Queen's University Press.

Lowe, R. (2007). *Introduction à la psychomécanique du langage*. Québec City: Les Presses de l'Université Laval.

Martone, A. (2006). La psicomeccanica di Gustave Guillaume. Una filosofia della mente? In S. Gensini & A. Martone (Eds.), *Il linguaggio. Teoria e storie delle teorie: in onore di Lia Formigari* (pp. 245–266), Napoli: Liguori.

Moignet, G. (1981). *Systématique de la langue française*. Paris: Klincksieck.

Pititto, R. (2012). *Pensare, parlare e fare. Introduzione alla filosofia del linguaggio*. Pomigliano d'Arco: Diogene Edizioni.

Soutet, O. (1998). *Manuale di linguistica generale*. (Trans. P. Giuliano). Bologna: il Mulino.

Stecher von, P. (2012). La lingüística de Gustave Guillaume. *De la lengua al discurso. OnOmázein, 5*(1), 163–180.

Wilmet, M. (1978). *Gustave Guillaume et son école linguistique*. Paris-Brussels: Nathan-Labor.

Time and Relativity of Time in Einstein's Theory of Special Relativity

Salvatore Principe

Abstract In 1905 Albert Einstein, in a paper entitled "*On the Electrodynamics of Moving Bodies*", as a solution to the disagreement between classical mechanics and the results of the Michelson's experiment, who showed the invariance of the speed of light in vacuum measured in different inertial reference systems, developed the theory of special relativity. In this essay Einstein expounded a theory that, instead of introducing a privileged system, required the revision of the concepts of space and time of classical physics. Combining the principle of Galilean relativity, according to which the laws of physics are invariant in all inertial reference systems, with the physics of electromagnetism, according to which the speed of light in a vacuum is constant, Einstein concluded that time is no more than a relative measure, namely that whenever we have to do with speed equal to or close to that of light, time is no longer a variable absolute and independent of the reference system adopted, but depends on the variable position. This is what Einstein shows through the critical examination of the concept of simultaneity. The abandonment of the traditional conception of space and time based on the idea of a spatial continuum flowing through a temporal continuum coherently leads to the assumption of a space-time continuum (chronotope) in which distances and time intervals vary with the changing the reference system, and together vary, of course, all other sizes to those connected (speed, acceleration, mass).

For the traditional mechanics all our mathematical and physical notions are based on the idea that a time interval and a space interval between two phenomena are always the same for any observer and in all conditions of observation. As is known, during the dispute on the Copernican theory, Galileo had enunciated the "principle of mechanical relativity" or the "principle of relativity of motion": two observers, one at rest and the other moving at a constant speed, see any mechanical phenomenon in the same way; for both of them, the mechanical behavior of bodies is identical. According to this principle the laws valid in a system at rest also apply to

S. Principe (✉)
Philosophy Section, Department of Humanities, University of Naples Federico II,
via Porta di Massa 1, 80133 Naples, Italy
e-mail: principesalvatore@gmail.com

F. Santoianni (ed.), *The Concept of Time in Early Twentieth-Century Philosophy*,
Studies in Applied Philosophy, Epistemology and Rational Ethics 24,
DOI 10.1007/978-3-319-24895-0_26

a system in uniform rectilinear motion. Galileo had expressed this "invariant" law through the "transformation formulas": $x - vt = x'$ $y = y'$ $z = z'$ $t = t'$, where the symbols x, y, and z are the spatial coordinates and t the temporal coordinate. From them it follows that there is no mechanical experiment to determine whether an inertial reference system is in a state of rest or in a state of uniform motion. Applying the principle of relativity to Copernicus' theory, Galileo was able to argue that it is not possible, with mathematical experience, to say that the Earth is stationary. It also follows that by comparing two supposed systems, one in quietness and the other in uniform motion, it can only be inferred that they move relative to one another, without however indicating which of the two is stationary and that can, therefore, be regarded as absolute reference system. It is not possible, therefore, to seize any system that is in rest or in absolute motion, from which to measure all other systems. Therefore, it is impossible to place in an objective way any phenomenon in space and time. Newton had thought he could get around the obstacle by postulating the absolute stillness of space and the absolute constancy of time and therefore the existence of an "absolute time" flowing evenly without reference to any external object, and an "absolute space". In Newton's theory the very notion of the relative seemed to be essential an absolute term of relationship; time and space, in fact, raised to the dignity of "sensors of God", were in fact the absolute conditions of all physical relationships. Post-Newtonian celestial mechanics, until Einstein, has been faithful to these assumptions, although not to the theistic illustration of them. Whether it accepted as a constant the fixed position of the stars, or postulated a motionless ether, it always needed to identify a constant invariant. The classical principle of relativity had thus unconditional validity as long what were being studied were slow movements such as the movement of the stars. But the examination of extremely fast movements, such as electricity and light gradually showed the insufficiency (not the contradiction) of the principle and made us observe the significant differences in the behavior of these phenomena. In 1881, in fact, on the basis of the principle of Galilean relativity, the Polish physicist Albert Michelson[1] believed to be able to measure precisely the speed of the rotation of the Earth by the instrumental measurement of the speed difference between two beams of light (Michelson 1881; Michelson & Morley 1886, 1887b, 1889): one of which was transmitted in the direction of the movement of the earth and the other in the opposite direction. According to the principles of classical physics, in accordance with the Galilean law of the composition of speeds, the speed of the beam which is shot in the direction of the motion of the earth should have been greater than that of the other beam shot in the opposite direction. Calculating the difference between the speeds of the two beams of light, would establish with precision the speed of the motion of the Earth. But the experiment's result was entirely the opposite of what

[1]Albert Abraham Michelson (December 19, 1852 in Strzelno, Province of Posen in the Prussian Partition—May 9, 1931 Pasadena, California) was an American physicist known for his work on the measurement of the speed of light and especially for the Michelson–Morley experiment. In 1907, he received the Nobel Prize in Physics. He became the first American to receive the Nobel Prize in sciences (Michelson 1881; Michelson & Morley 1887a).

everyone expected. The two light rays were shot with the same constant speed of 300,000 km/s. The result of the experiment thus posed a new problem: how to explain the constancy of the speed of light and the independence of this speed from the state of rest or motion of the observer or the light source. The "accidental" discovery of the constancy of lightspeed put the scientists in front of the fact that in nature only the relative motion would have meant: there is no fixed background of points in space, within which to measure in absolute terms the speed and displacement of a body; and there is no absolute flow of time, within which to measure time intervals equally valid for any observer. The concepts of space and time and therefore speed, which is the relationship between space and time are not absolute values, but closely related to the observer that measures them. The ideas of a length and an absolute space, a time and a flow of absolute time, are metaphysical concepts that go beyond what the experience and the observer can justify.

In 1905 Albert Einstein, in a paper entitled "*On the Electrodynamics of Moving Bodies*"[2], as a solution to the disagreement between classical mechanics and the results of Michelson's experiment, which showed the invariance of the speed of light in vacuum measured in different inertial reference systems, developed the theory of special relativity. In this essay Einstein expounded a theory that, instead of introducing a privileged system, required the revision of the concepts of space and time of classical physics. Einstein started from a hypothesis which rendered "relative" the Newtonian absolutes and replaced these with a new "absolute": the speed of light, not only considered as "constant", but as "speed-limit", and for this reason "absolute". Combining the principle of Galilean relativity, according to which the laws of physics are invariant in all inertial reference systems, with the physics of electromagnetism, according to which the speed of light in a vacuum is constant, Einstein concluded that time is no more than a relative measure, namely that whenever we have to do with speed equal to or close to that of light, time is no longer a absolute variable independent of the reference system adopted, but one that depends on the variable position. This is what Einstein showed through the critical examination of the concept of simultaneity. Einstein's argument is roughly as follows: two events, for example, two light signals emitted when a chronometer marks noon by two light sources positioned at the two ends *A* and *B* of a straight stretch of road, will be simultaneous for an observer *O* who stay still in the median point *M* of the *AB* axis or perpendicular to that axis at the point *M*. But for another observer *O'* that moves along the axis or in the direction of *A* or in the direction of *B*, the two events will not appear simultaneously; *A* will be earlier than *B* or vice versa, depending on the direction of motion of the same observer. Therefore, the simultaneity and the order of succession of events (which is what is meant by time) are defined only in relation to the reference system in which the observer takes the measures, and thus are relative magnitudes.

[2]Einstein (1905).

In the popular explanation of relativity published in 1917[3], Einstein showed the relativity of simultaneity through the well-known experiment of the two lightning strikes on a rail at a certain distance from each other. They will be simultaneous when an observer at the midpoint between the two lightning records the glow of the two lightning at the same time, and this presupposes that the light travels along the two paths at the same speed. Now suppose a very long train travels along the track at speeds *V*. An observer who is on the train applies the same procedure for the detection of simultaneity used by an observer on the ground, with the difference that, instead of the track, he will use the train as a rigid body of reference. We call *M* the midpoint between the two points at which the lightning bolts fall on the rail and we place an observer at point *M′* on the train aligned exactly with *M* when the two lightning bolts fall, measuring the time from the point of view of the observer on the ground. Now, says Einstein, if the simultaneity is the same in all reference systems, then the observer on the train should register the two glows at the same time; instead, as the train travels in the direction of one of the lightning bolt and away from the other, the observer will first perceive the glow that comes from the front of the train. Therefore, the simultaneity thus defined is different on the train and on the ground, or is relative to the state of motion of the reference system.

Similarly, it is also worthwhile referring to the absoluteness of the unit of measurement (ruler or clock) as something absolute on which to base the absoluteness of time. In fact as the simultaneity is belied by the previous mental experiment, there is equally refuted the synchrony of clocks of observers compared to the simultaneity of events. Each unit of measure applies relatively to each system, and for different systems, for example by varying the speed; the unit of measurement gets longer or shortens correlatively. Each reference system has its own measure of spatial distances and time intervals; and the measurements are not the same when those systems are different. The space measured by an observer at rest differs from what is measured by a moving observer; and the same applies to the time.

The abandonment of the traditional conception of space and time based on the idea of a spatial continuum flowing through a temporal continuum coherently leads to the assumption of a space-time continuum (chronotope) in which distances and time intervals vary with the changing of the reference system, and together there vary, of course, all other sizes to those connected (speed, acceleration, mass). Time therefore becomes a fourth dimension homogeneous with the three spatial dimensions. Hence, the measurements of temporal intervals and spatial lengths, carried out by inertial observers do not necessarily correspond with each other, giving rise to phenomena such as the dilation of time and the contraction of the lengths. The maximum limit of the spatial contraction and temporal expansion is the speed of light.

Such spatiotemporal variations that are poorly or not at all noticeable in slow movements such as those that we experience on a daily basis, however, become of

[3]Einstein (1917)

considerable magnitude at gradually increasing speeds in the direction of the speed of light. A somewhat paradoxical example of temporal dilation due to extremely fast motion is the so-called experiment of the twins. Imagine a pair of twins, Anne and Betty. The latter departs with a spacecraft moving at a speed close to that of light and returns to Earth after a few years. Anne instead remains on Earth (and moves with the Earth, which has a speed of little relevance). When Betty returns, she will find Anne older.

Suppose hypothetically that Betty departs in 2000 and returns in 2020. For Ann the absence of her sister will last 20 years and then she will have grown old equally. If Betty had traveled at a speed of 240,000 km/s, according to Einstein's formula, in her reference system the trip should have lasted only 12 years. Betty will return in the year 2020 having actually lived for 12 years and having grown only 12 years old. She will remain surprised by the fact that twenty Earth years have elapsed, which for her lasted only 12; however, the aging of her sister will testify to this.

It's important to understand two things. First, the effect of the twins is real, not just a mental experiment. Second, it has nothing to do with the effect of motion on the aging process. We should not think that the years spent in the spacecraft are somewhat more lenient towards Betty because of her confinement or moving through space. The best way to examine the experiment of the twins is in terms of events. There are two events limit: Betty's departure from Earth and her subsequent return. The twins must agree on the time when these events occur, because they are both witnesses. The problem is therefore that for Ann 20 years separate the two events, while for Betty 12 years separate them. With an apparent contradiction, both observe the one in the system of the other: the de-synchronization of clocks, dilation of time, and therefore the slowdown of all the movements that occur in the reference frame in motion, length contraction in the sense of motion.

So different observers spend different time intervals between the same events. Anne's time and the Betty's time and are not the same thing. There is not a constant time interval between two events, nor a real duration, but only relative time differences.[4]

References

Bergmann, G. P. (1976). *Introduction to the theory of relativity*. New-York: Dover Publications.

Cassirer, E. (1921). *Zur Einsteinschen Relativitätstheorie*. Erkenntnistheoretische Betrachtungen, Berlin: Bruno Cassirer Verlag.

Eddington, S. A. (1920). *Space, time and gravitation: an outline of the general relativity theory*. Cambridge: Cambridge University Press.

[4]About the issues of the relativity of time see also Einstein (1905), Einstein (1917), Einstein (1922), Bergmann (1976), Cassirer (1921), Eddington (1920), Einstein (1922), Lavenda (2011), Ludyk (2013), Russell (1969), Hawking and Mlodinow (2005), Pauli (1981), Reichenbach (1928), Schlick (1915, 1917), Sexl and Schimdt (1978).

Einstein, A. (1905). Zur Elektrodynamik bewegter Körper. *Annalen der Physik, 17*, 891–921. Republished in *Das Relativitätsprinzip. Eine Sammlung von Abhandlungen*, pp. 27–52, Blumenthal, Otto, ed. Leipzig: Teubner, 1913, 27–52; trans. engl. by A. Beck, *On the Electrodynamics of Moving Bodies, in The Collected Papers of Albert Einstein* Vol. 2, edited by John Stachel, David C. Cassidy, Jürgen Renn, and Robert Schulmann, Doc. 23, 275–310; trans. 140–171, Princeton University Press. Available from http://einsteinpapers.press.princeton.edu/vol2-trans/154.

Einstein, A. (1917). *Über die spezielle und die allgemeine Relativitätstheorie (Gemeinverständlich)*. Braunschweig: Vieweg, trans. by R. W. Lawson & A. Engel, *On the Special and General Theory of Relativity (A Popular Account)*, in *The Collected Papers of Albert Einstein*, Vol. 6 edited by Martin J. Klein, A. J. Kox, and Robert Schulman, Doc. 42, 420–539; trans. 247–420, Princeton University Press. Available from http://einsteinpapers.press.princeton.edu/vol6-trans/259.

Einstein, A. (1922). *The meaning of relativity*. Princeton: Princeton University Press.

Hawking, S., & Mlodinow, L. (2005). *A briefer history of time*. New York: Bantam Dell.

Lavenda, H. B. (2011). *A new perspective on relativity: an odyssey in non-euclidean geometries*. Singapore: World Scientific.

Ludyk, G. (2013). *Einstein in matrix form*. Berlin: Springer.

Michelson, A. A.,& Morley, E. W. (1886). Influence of motion of the medium on the velocity of light. *American Journal of Science, 31*, 377–386.

Michelson, A. A., & Morley, E. W. (1887a). On the relative motion of the earth and the luminiferous ether. *American Journal of Science, 34*, 333–345.

Michelson, A. A., & Morley, E. W. (1887b). On a method of making the wave-length of sodium light the actual and practical standard of length. *American Journal of Science, 34*, 427–430.

Michelson, A. A., & Morley, E. W. (1889). On the feasibility of establishing a light-wave as the ultimate standard of length. *American Journal of Science, 38*, 181–186.

Michelson, A. A. (1881). The relative motion of the Earth and the luminiferous ether. *American Journal of Science, 22*, 120–129.

Pauli, W. (1981). *Theory of relativity*. New York: Dover Publications.

Reichenbach, H. (1928). *Philosophie der Raum-Zeit-Lehre*. Berlin: Walter de Gruyter.

Russell, B. (1969). *The ABC of relativity*. London: Allen & Unwin.

Schlick, M. (1915). Die philosophische Bedeutung des Relativitätsprinzips. *Zeitschrift für Philosophie und philosophische Kritik, 159*, 129–175.

Schlick, M. (1917). *Raum und Zeit in der gegenwärtigen Physik*. Berlin: Verlag von Julius Springer.

Sexl, R., & Schimdt, K. H. (1978). *Raum-Zeit-Relativität*. Amburg: Rowohlt.

Tenses and Temporality in Reichenbach's Thought

Michele Malatesta

Abstract *Elements of Symbolic Logic* by Hans Reichenbach provides the first analysis of tenses from a translinguistic, transgrammatical, transcultural logical viewpoint. However, the author does not address the problem of relations between the tenses and temporality, whose investigation is devoted to physics. Despite the brilliant discovery, an irreconcilable gap between the two different epistemological perspectives emerges at a glance. Where Reichenbach stopped is precisely where one must start from to continue the exploration on the exterminated continent that one can glimpse behind his study

1 Preliminary Aspects: Logic, Psychology, and Language

Generally, the authors of manuals of logic enter immediately *in medias res* (Shoenfield 1967; Barnes and Mack 1975) or place a chapter on semiotics before sentence calculus (Carnap 1942; Malatesta 1997). Reichenbach instead starts from the relationship between logic and psychology before facing the problem of signs. Logic "deals with the laws of thought" (Reichenbach 1947: 1) begins the philosopher, who doubtless will have had in mind the mature work of George Boole.[1] Psychological laws must be distinguished from logical laws: logic takes

[1]«The design of the following treatise is to investigate the fundamental laws of those operations of the mind by which reasoning is performed» (Boole 1854: 1). See on this topic the incisive essay of Nuzzetti (1986).

M. Malatesta (✉)
Philosophy Section, Department of Humanities, University of Naples Federico II, via Porta di Massa 1, 80133 Naples, Italy
e-mail: malatesta@mclink.it

F. Santoianni (ed.), *The Concept of Time in Early Twentieth-Century Philosophy*, Studies in Applied Philosophy, Epistemology and Rational Ethics 24,
DOI 10.1007/978-3-319-24895-0_27

care of the latter. On this point, there is a perfect convergence between Reichenbach and Frege.[2]

The mental process is irregular: partly logical and partly automatic, to which is not extraneous the emotive component. According to Reichenbach, the observable laws formulated in psychology include correct and erroneous thought, inasmuch as the tendency to commit certain fallacies must be considered a psychological law in the same sense of the most fortunate habits of correct thought.

If one wishes to say that logic has to do with thought, one must specify that

> logic teaches us how thinking *should* proceed and not how it does proceed (Reichenbach 1947: 1)[3]

but the philosopher hastens to clarify that creative thought processes do not move along prepared paths, but follow a method of trial and error. Creative thought follows its often obscure and often unconscious paths.

There should be distinguished two realms of analysis that can be called the context of discovery and the context of justification. The former is left to psychological analysis, while logic addresses the latter. The second process is linked to language (Reichenbach 1947: 2). Only after the processes of thought have been expressed in linguistic form, they do reach the precision that makes them accessible to logical tests. In the theory of deduction, we study the rules that lead from true linguistic expressions to true linguistic expressions.

The method of symbolization has proven to be a useful instrument for the clarification of language. The great advantage of modern logic over ancient forms of logic consists in the fact that this logic is capable of analyzing structures, which traditional logic was incapable of analyzing, and of resolving problems, whose existence the latter did not even suspect.[4]

2 The Analysis of Language

If logic is the analysis of language, logical inquiry must take its moves from the examination of language. Language consists of signs and signs are physical things: traces of ink on paper, traces of chalk on the blackboard, and waves produced by

---

[2]«[…] the laws of logic can be called 'laws of thought': so far as they stipulate the way in which one ought to think […] But the expression 'laws of thought' leads one to suppose that these laws govern thinking in the same way as laws of nature govern events in the external world. In that case they can be nothing but laws of psychology: for thinking is a psyche process. And if logic were concerned with these psychological laws it would be part of psychology […] " (Frege 1893).

[3]«It is curious that the anti-Kantian Reichenbach share the same opinion on this topic as the philosopher from Königsberg: "The question of logic is not… how we think, but how we should think"» (Kant 1902: 14).

[4]For a more in-depth treatment, see also Reichenbach (1951: 215–229).

the human throat. What makes these physical things *signs* is "is the intermediary position they occupy between an object and a sign user, i.e. a person" (Reichenbach 1947: 4). The philosopher makes a further distinction between indexical, iconic, and conventional signs.

Indexical signs are phenomena that refer to other phenomena, e.g., smoke is a sign of fire[5]; *iconic signs* are representations that preserve a certain similarity with the objects they denote, e.g., photographs, geographical maps, cadastral maps, all scale reproductions; *conventional signs* or *symbols* are expedients devised to refer to objects without any similarity with them, e.g., the different words used by the various natural languages to refer to the same objects. The threefold distinction dates back to Pierce as Reichenbach explicitly recognizes (Reichenbach 1947: 4).

Signs are not watertight compartments but are united one to the other. The most important among the signs is the proposition. While analytic philosophy makes a distinction between sentence, proposition, and statement,[6] Reichenbach uses these terms interchangeably. He writes

> We do not distinguish between 'proposition', 'sentence', and 'statement', and shall therefore use these terms interchangeably (Reichenbach 1947: 5).

What makes of a proposition a fundamental unit is its truth value, that is, the fact that it can be true or false (Reichenbach 1947: 6). An isolated word is neither true nor false: "the property of having a *meaning* is originally restricted to whole sentence" (Reichenbach 1947: 6). If we sometimes speak of the meaning of a word, we mean to say that we understand the meaning of a word if we know how to use it in sentences with different meanings. The result is that «sentence-meaning is logically prior to word-meaning, i.e., that the expression 'word-meaning' is defined in terms of the expression 'sentence-meaning'» (Reichenbach 1947: 6). Here, Reichenbach clearly takes up a position against the principle of compositionality of Frege,[7]

[5]Reichenbach makes a different use of the expression "*indexical signs*" than that made by linguists. For the latter, *indexical* or *deictic* signs are those signs whose interpretation varies with the contest, like the use of personal pronouns (ex. "I" spoken by person A refers to A; pronounced by person B, it refers to B) or the use of demonstrative adjectives and pronouns (ex. "this" refers to an object close to A, if A is speaking, and to an object close to B, if B is speaking), etc.

[6]The *sentence* is a linguistic expression susceptible to being true or false; the *proposition* is the meaning of a sentence; the *assertion* is a sentence used to affirm with certainty.

[7]"The names, whether simple or themselves composite, of which the name of a truth-value consists, contribute to the expression of the thought, and this contribution of the individual [component] is its *sense*. If a name is part of the name of a truth-value, then the sense of the former name is part of the thought expressed by the latter name" (Frege 1893: 90). A young scholar (Tripodi 2008) retains that the point of view of the *Grundgesestze* (1893–1903), which are posterior to *Über Sinn und Bedeutung* (1892), overcomes the conception expressed by Frege in *Die Grundlagen der Arithmetik* (1884) where the German logician had clearly said that "only in the context of a statement do the words mean something" (Frege 1884: § 62); moreover, he had added "one must never investigate the meaning of a word in isolation" (Frege 1884: ix).

a principle that is found again in Wittgenstein.[8] Doubtless Frege and Wittgenstein are right for what concerns mathematical language, but cannot make of the principle of compositionality a linguistic universal. Reichenbach, who taught in Turkey, knows well what happens in an Altaic language. Turkish, which uses many suffixes, sometimes expresses a whole statement with a single word. For example, the word '*alabīleceğim*' is a single sentence and means 'I shall be able to buy' (Reichenbach 1947: 5). It is a truly difficult undertaking to try to deduce the meaning of the whole from the individual parts: the meaning of the individual parts is understood starting from the meaning of the whole.

3 Linguistic Levels and Semiotics

Signs are physical things coordinated with other things on the basis of rules. The process of coordination can be repeated; in this manner, signs can be introduced that refer to signs (Reichenbach 1947: 9). The repetition of the coordination of signs was never invented by logicians. Already in ordinary language there are many terms of this kind: the word 'word' refers to signs; just as the words 'sentence,' 'proposition,' 'phrase,' and 'name.' Signs of signs constitute a higher level that is called *metalanguage*, while the ordinary language is called *object language.* We can create other signs that refer to signs of signs. We will have, in this manner, a *metametalanguage* (Reichenbach 1947: 9).

Reichenbach considers physical objects to be at level zero. "Physical objects divide into *things*, such as individual human beings, tables, atoms, and *situations*, also called *states of affairs*, which constitute the denotata of sentences. Thus the sentence 'the battleship *Bismarck* was sunk' denotes a situation; the ship itself is a thing" (Reichenbach 1947: 14–15). This is an important step that marks Reichenbach's detachment from Frege,[9] Church,[10] and Carnap,[11] for whom the denotation of a sentence is the truth. To illustrate his viewpoint, the German philosopher creates an interesting table.[12]

[8]«Man versteht ihn (*sc.* einen Satz), wenn man seine Bestandteile versteht» (Wittgenstein 1961: 4.24).

[9]"We are pushed then to recognized the truth value of a phrase as its denotation" Frege 1892: 34).

[10]"And we declare all true sentences to denote the truth-value truth, and all false sentences to denote the truth-value falseness" (Church 1956: 25).

[11]"By intension of a sentence we mean the proposition that it deisgnates, and by extension its truth value" "(*Unter der Intension eines Satzes wollen wir die durch ihn bezeichnete Proposition, und unter seine Extension seinen Wahrheitswert*)" (Carnap 1954: 40).

[12]A propositional variable is a sign that stands for a proposition at will.

Objects	*Object language*	*Metalanguage*
	bird thing	'bird' word, name
	the bird flies a_1 situation	'the bird flies' 'a_1' sentence, proposition
	a	'a' propositional variable
		'the bird flies' is true α_1
		α

Reichenbach returns to his refusal to distinguish between sentence and proposition. He writes "When some logicians thought it necessary to distinguish between 'proposition' and 'sentence' they did so because they believed that there was a third thing between the sentence, i.e. the linguistic expression, and the situation. Such a third thing is certainly unnecessary, and we shall therefore identify sentence and proposition" (Reichenbach 1947: 15). It is strange that a thinker like Reichenbach, who made an impact with a non-indoeuropean language like Turkish, and who denied the principle of compositionality, does not realize that this third thing is necessary to translate a language belonging to a radically different linguistic group, with a radically different grammar and forms of speech.

As for the general theory of signs, the philosopher accepts the tripartite division of semiosis, but, unlike Carnap who starts from more complex relations to arrive at more simple ones (pragmatic, semantic, syntax) (Carnap 1942: 8–15), Reichenbach, following the father of semiotics (Morris 1938: 37–111), goes from the simpler to the more complex (syntax, semantic, pragmatic). He adds in lapidary fashion: "The third part, *pragmatics*, adds a reference to persons; it therefore refers to things, signs, and persons" (Reichenbach 1947: 15–16).

4 The Tenses of Verbs

After having expounded the whole of classical logic without first examining syntax and then semantics, Reichenbach dwells at length on conversational language. And it is in this field that the philosopher carries out fundamental contributions, turning over a terrain that had never been plowed, or better, discovering worlds never before taken into consideration by modern logic. Reichenbach's study marks the point of departure for some new fields of logic that have begun to formalize various sectors of everyday language. Even if the German philosopher cannot be numbered among the founders of the various branches of heterodox logic, however, without his reflections, many of these would never have been born.[13] I will here take into consideration only the tenses of verbs.

Never had so rigorous an analysis of the tenses of verbs been done before Reichenbach. The philosopher starts from the linguistic act that he calls the 'point of speech,' with respect to which the three statements 'before the point of speech,' 'simultaneous with the point of speech,' and 'after the point of speech' furnish only three verb tenses. However, the structure of verbal tenses is much more complex. "From a sentence like 'Peter had gone' we see that the time order expressed in the tense does not concern one event, but two events, whose positions are determined with respect to the point of speech. We shall call these time points the *point of event* and the *point of reference*. In the example the point of the event is the time when Peter went; the point of reference is a time between this point and the point of speech" (Reichenbach 1947: 288).

Reichenbach focuses on five languages: English, Turkish, classical Greek, French, and German. He symbolizes then with '*E*' the point of the event, with '*R*' the point of reference, with '*S*' the point of the linguistic act, and with the arrow, the direction of the time.

It is thus easy for him to analyze the tenses of verbs that refer to precise events whether past, present, or future, giving them a time relative to the graphic representation in the English language (Reichenbach 1947: 290).

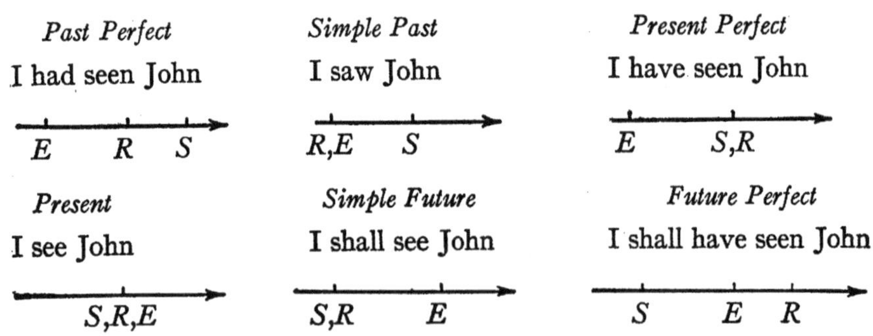

[13]Think of the logic of deitics, the logic of assertions, the erotetic logic, and the logic of commands, which will be born from the considerations which Reichenbach makes, respectively, in Reichenbach (1947): 284–287, 336–339, 339–342, 342–343.

Moreover, because the English language uses the present participle to indicate that the event covers a certain length of time, Reichenbach delineates another chart (Reichenbach 1947: 290).

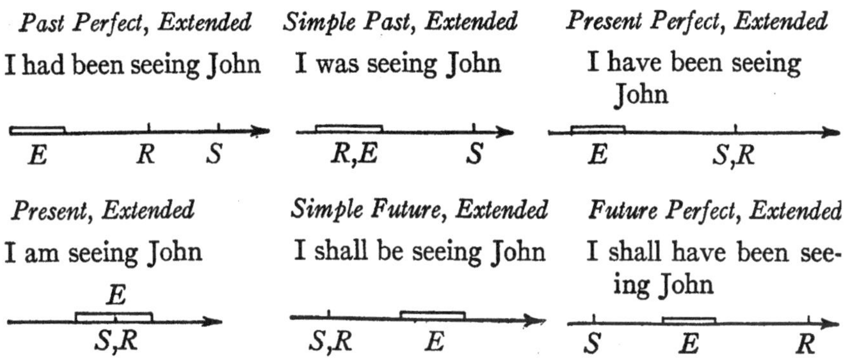

The extended times are sometimes used to indicate not the duration of an event, but its repetition. Reichenbach does not say it but his observation regards those verbs that grammarians call *frequentative verbs*, that is, those verbs that express a repeated action and that they abound in the Latin language.[14] The philosopher observes that while English expresses extended tenses of verbs—whether past, present or future—using the present participle, other languages have developed special suffixes to reach the same goal. Turkish possesses a tense of this kind, called *muzari*, which indicates repetition or duration of an event that goes from the past to the future through the present. An example of this verb tense is the word 'görürüm,' which can be translated with the sentence 'I usually see.'

The syllable 'gör' is the root that means 'to see'; 'ür' is the suffix that expresses the muzari; 'üm' is the suffix that expresses the first person 'I.' The statement 'I see' corresponds to the Turkish 'görüyorum': the difference with the previous sentence is given by the infix 'üyor,' at the center of the word, which expresses the present tense.

Classical Greek uses the *aorist* to express the repetition or habitual recurrence in the present tense. The aorist, however, is originally a non-extended past tense, and has taken on the second use by a shift of meaning. The aorist understood as extended time is called *gnomic aorist.* Following William Watson Goodwin, the famous Greek scholar who taught at Harvard (Goodwin 1930: 275), Reichenbach explains in this manner the shift of meaning: once having established a typical case in the past, the listener is left with the inductive inference that under similar

[14]For example, the frequentative of 'ago' *agire* [to act] is 'agito' *to agitate, shake, push here and there*; the frequentative of 'rogo' *chiedere* [to ask] is 'rogito' *to ask repeatedly, with insistence*; the frequentative of 'teneo' *tenere* [to have] is 'tento' *to touch, handle*, etc.

conditions, the same thing will be repeated in the future is left to the listener. Such a shift of meaning is furnished by the English adage 'Faint heart never won fair lady' (Reichenbach 1947: 291).[15]

French and German do not have extended verb tenses: to reach that objective, they use special words such as 'always,' 'usually,' etc. In French, however, we find an exception for what concerns the past. In this language, there are two different verb tenses: (a) the *imperfait* (e.g., "je voiais Jean"), which corresponds to the English simple past-extended (e.g., "I was seeing John"); (b) the *passé défini* (e.g., "je vis Jean"), which corresponds to the English *simple past* (e.g., "I saw John").[16] The same distinction is found in classical Greek: in this language, the imperfect corresponds to the French imperfait and to the English simple past-extended, while the aorist, in its original meaning of past tense, corresponds to the French passé défini and, in consequence, to the English simple past.

5 From the Logic of Tenses of Verbs to the Logic of Temporality

Reichenbach's most original contribution consists in the attempt to rationalize verb tenses regardless of the verb tenses present in the various languages, which often arise rhapsodically on different or even diverging bases (consider the difference, from the grammatical point of view, between "io vedrò," "ich werde sehen," and "I shall see" which are explained on the basis of three different intentional perspectives which are thus relevant from a pragmatic point of view.[17]

The author distinguishes 13 possibilities. He chooses the point of speech as the point of departure. He indicates the position *R* with respect to *S* with the terms 'past,' 'present,' and 'future': in this manner, one obtains already three possibilities. He then indicates the position of *E* with respect to *R* with the words 'anterior,' 'simple,' and 'posterior,' where the word 'simple' denotes the coincidence of *R* and *E*: in this way, one obtains $3 \cdot 3 = 9$ forms, called *fundamental forms*. Further differences of form are obtained if one takes into consideration the position of the event relative to the point of speech. For example, the form *S–E–R* can be

[15]I like to underline that such a shift of meaning is found also in some Italian sayings: '*donna baffuta è sempre piaciuta*' ['A bewhiskered woman always pleased'] '*caldo di panni non ha fatto mai danni*' ['Heat from clothing was never harmful,' meaning that in the cold, the heat procured from clothing is helpful, while when it is hot, one can simply take off the clothing—Trans.m], etc.

[16]Also, in Italian there is an analogous difference between the *imperfetto*, (ex. "*io vedevo Giovanni*", "I was seeing John") corresponding to the French imperfait and to the English simple past-extended and the *passato remoto* (ex. "*io vidi Giovanni*", "I saw John") corresponding to the French passé défini and to the English simple past.

[17]Think of the anomaly of the Italian language that distinguishes full four forms of past (*passato prossimo, imperfetto, passato remoto, trapassato remoto*) against the two of the futhre (futuro anteriore, futuro semplice), while coherence would have a symmetry between the verb tenses of the past and those of the future.

Structure	New name	Traditional name	
E–R–S	Anterior past	Past perfect	
E, R–S	Simple past	Simple past	
R–E–S R–S, E R–S–E	}		
		Posterior past	–
E–S, R	Anterior present	Present perfect	
S, R, E	Simple present	Present	
S, R–E	Posterior present	Simple future	
S–E–R S, E–R E–S–R	}		
		Anterior future	Future perfect
S–R, E	Simple future	Simple future	
S–R–E	Posterior future	–	

distinguished from the form *S, E–R* with respect to the relation between *S* and *R* on the one hand and *R* and *S* on the other; however, these two forms are not different and can be considered a single form, and so on. Therefore, the 13 forms can be reduced to the nine fundamental ones.

Reichenbach, though having glimpsed the possibility of a logic of temporality underlying the logic of verb tenses, at this point stops. The fact that the English language lacks a tense for the posterior past and a tense for the future posterior, while these verb tenses exist are foreseen by the logical calculation, should have opened his eyes. Those verb tenses are lacking in English but not the capacity of Anglophones to express the reference to an event that takes place in a posterior past with respect to the simple past or which takes place in a posterior future with respect to the simple future. Reichenbach, given his cultural formation, remained blocked before the new unexplored continent, and retained that the inquiry into temporality, unlike that into verb tenses, was the exclusive competence of physics and not of logic and of the relative ontological implications.[18]

6 Reichenbach's Legacy

Reichenbach's study initiates a vast flourishing of two types of research, concerning, respectively, tense logic and temporal logic. The awakening, however, occurs after some delay: *Elements of Symbolic Logic* precede by 10 years Prior's *Time and Modality* (1957), by nineteen Cocchiarella's *Tense Logic* (1966), by twenty Prior's *Past, Present and Future* (1967), by twenty-one *Papers on Time and Tense* by the same author (Prior 1968), and by a full 24 years *Temporal Logic* by

[18]See, among the numerous works by Reichenbach on time, "What is Time?" (Reichenbach 1951: 144–156).

Rescher and Urquhart, even if this latter work constitutes the sum of all the works produced after Reichenbach's original study (Rescher and Urquhart 1971).[19] However, in my way of seeing things, these scholars have not come into Reichenbach's legacy. They have only taken inspiration from the philosopher from Hamburg, but for following different paths than those espied by him, even if not followed by him.

7 Beyond Reichenbach Through Reichenbach

Reichenbach's brilliant perspective, which can be shared by someone who like me is an impenitent *homo metaphysicus* (and thus radically antineopositivist), has a limit: while the negation of the principle of compositionality and the negation of the conception that denoted by sentences is a truth value have allowed the German philosopher to take giant steps, the refusal to distinguish between sentence (understood as a linguistic expression) and proposition (conceived as the meaning of the sentence) have not allowed him to make the last jerk to cross the finish line: the passage from verb tenses to the condition that founds them. If one makes the meaning coincide with the statement exactly, Reichenbach cannot justify why the Chinese statement

> "Shūzhuō shàng yǒu liǎng běn shū"
> (lit. *Desk above to have two volume book*)

is translated into English with the statement

> "There are two books on the desk".

If the translation is possible, it means that a single thought is immanent in two different grammatical structures and therefore does not coincide *sic et simpliciter* with any of the two. In fact, if sense and grammatical form were to coincide exactly with only one of the two grammatical structures, the same sense could not be expressed by the other. If this distinction is not made, one does not understand anything when dealing with languages that have just two verb tenses, like the Semitic languages that have only the perfect and imperfect. Perhaps those who speak Hebrew, Arabic, Amharic, or Tigrigna are unable to express the anterior or posterior past with respect to a point in the past, and the anterior or posterior future with respect to a point in the future. In these cases, the corresponding verb tenses are lacking but not the linguistic expressions to express their meaning. And what to say about languages that do not have verb tenses other than the present, a verb tense which, moreover, cannot be conjugated, like Chinese?

---

[19]See the very interesting bibliography, exposed first in chronological order (259–264), and then by author (264–267).

Where Reichenbach stopped is precisely where one must start from to continue the exploration on the exterminated continent that one can glimpse behind his study. Without rejecting anything of Reichenbach's analyses, it is necessary to climb onto his shoulders, that is, one must go back from natural languages to the logical structure of temporality, which is a universal structure, transgrammatical, and transcultural, a true Mendeleev's table of temporality, constructed with the combinatory calculus of mathematics, on which there move not only the natural languages currently spoken, whether inflected or agglutinating or isolating, but also those which have disappeared and those never spoken but logically possible (Malatesta 1992).

References

Barnes, D., & Mack, M. J. (1975). *An algebraic introduction to mathematical logic*. New York, Heidelberg, Berlin: Springer.

Boole., G. (1854). *An investigation of the laws of thought on which are founded the mathematical theories of logic and probabilities*. London: MacMillan [Repr:].

Carnap., R. (1942). *Introduction to semantics*, Cambridge, London: Harvard University Press [Repr. 1975].

Carnap, R. (1954). *Einführung in die symbolische Logik*. Wien: Springer.

Church, A. (1956). *Introduction to mathematical logic* (Vol. I). Princeton: Princeton University Press.

Cocchiarella., N. (1966). *Tense logic. A study of temporal reference* (Doctoral Dissertation). Los Angeles: University of California.

Frege, G. (1884). *Die Grundlagen der Arithmetik; eine logisch mathematische Untersuchung über den Begriff der Zahl (English edition)*. Breslau: Koebner.

Frege, G. (1892). Über Sinn und Bedeutung. *Zeitschrift für Philosophie und philosophische Kritik, 100*(1), 25–50.

Frege, G. (1893). *Grundgesetze der Arithmetik, begriffsscriftlich abgeleitet (English edition)* (Vol. I). Jena: Hermann Pohle.

Goodwin, W. W. (1930). *Greek Grammar*. Boston: Ginn & Company.

Kant., I. (1902). Logik. Ein Handbuch zu Vorlesungen. In Kant, I. 1902. *Gesammelte Schriften*, ed. König Ludwig. Berlin: Akademie der Wissenchaften.

Malatesta., M. (1992). Logic space of temporality in natural languages. In F. Collot (Ed.), *13e Congrès International de Cybernétique. Actes Namur, Belgique, 24–28 Août 1992*, (pp. 136–140). Namur: Association Internationale de Cybernétique.

Malatesta, M. (1997). *The primary logic: Instruments for a dialog between two cultures*. Leominster (UK): Gracewing.

Morris, W. C. (1938). *Foundations of the theory of signs (Italian edition)*. Chicago: University of Chicago Press.

Nuzzetti, M. (1986). *Logica e linguaggio nella filosofia di George Boole*. Napoli: Liguori.

Prior, N. A. (1957). *Time and modality*. Oxford: Oxford University Press.

Prior, N. A. (1967). *Past, present and future*. Oxford: Oxford University Press.

Prior, N. A. (1968). *Papers on time and tense*. Oxford: Oxford University Press.

Nicholas, R., & Urquhart, A. (1971). *Temporal logic*. Berlin: Springer.

Reichenbach, H. (1947). *Elements of symbolic logic*. New York, London: MacMillan.

Reichenbach, H. (1951). *The rise of scientific philosophy*. Berkeley, Los Angeles: University of California Press.
Shoenfield, R. J. (1967). *Mathematical logic*. Reading, Menlo Park, London, Don Mills: Addison Wesley Publishing Company.
Tripodi, V. (2008). Contestualità e composizionalità del significato. *Humana Mente, 4*, 1–26.
Wittgenstein, L. (1961). *Tractatus Logico-Philosophicus*. London: Routledge & Kegan Paul.

The Concept of Time in Prigogine

Nicola Grana

Abstract What is time? What is its beginning, if we may speak of a beginning? All these questions are not stimulated by a crypto-metaphysical need, but by the epistemological approach itself. It is enough here to think of Ilya Prigogine, who has made of the concept of time the main task of his scientific and philosophical research. In this horizon, we must accept what Stephen Hawking himself said about the beginning and the end of time, in physic-cosmological meaning, as we can accept the meaning of experienced time, which stands in the evolution of our history, which begins with us, coincides with the origin of our biological time, or of our biological times, to end with the end of our biological history on a macroscopic scale. Our evolutionary history is, of course, underlined by experienced time, the charioteer of our changes, but in a dialectical relation with chronological, chronometric, chronosophic times, in a relation of one among many, which produces states of suffering. But it does not make this evolutionary history less interesting. So, at the end of our journey, at least the consciousness of being "inhabitants" of time and *bearers* of change appears, and this occurs whether we have behind us *big cataclysms* or thermic Death.

Naturally, the study of temporal order is not univocal, in fact there is a linear approach, and an approach with several branches going in different directions, such as the relativistic approach, a circular approach that allows you to go back to the starting point, and so on. Other approaches are of the discreet or continuous kind; in this case their compactness or density give rise to problems related to infinite divisibility. Furthermore, there is the approach involving a macro-aspect, related to the totality of time, going from a past across the present toward a future (and this affects the change of our states), and the approach involving a micro-aspect related to the interior structure made of limited intervals of time (Von Wright 1974: 271).

Do not all these approaches suggest dealing with the concept of time in a unique way? To clarify its logical, conceptual, factual nature, and so on? What is a change?

N. Grana (✉)
Philosophy Section, Department of Humanities, University of Naples Federico II, via Porta di Massa 1, 80133 Naples, Italy
e-mail: grana@unina.it

F. Santoianni (ed.), *The Concept of Time in Early Twentieth-Century Philosophy*, Studies in Applied Philosophy, Epistemology and Rational Ethics 24, DOI 10.1007/978-3-319-24895-0_28

> In its simplest way it is a transition from, or a transformation of, one state into another. We may call the first term of the relation «initial state», and the second «final state». The words «initial», «final» and «transition» suggest a temporal order[1] (Von Wright 1974: 259).

And so, what is time? What is its beginning, if we may speak of a beginning? All these questions are not stimulated by a crypto-metaphysical need, but by the epistemological approach itself. It is enough here to think of Ilya Prigogine, who has made of the concept of time the main task of his scientific and philosophical research.

Let us begin stressing the fact that according to Prigogine time leads us to man and not vice versa; man be not the creator of time. This position is totally different from what the physicist John Archibald Wheeler thinks: for him, man, the observer, his consciousness, creates time, which would not be there nor exists in a world without men and their consciousness. On the other hand, Ilya Prigogine thinks of man as being part of this flow of irreversibility which is one of the essential elements, consubstantial to the universe (Prigogine 1988: 21) and finds its essential role along the path followed by Henri Bergson. Time is a subject of science just because it has a fundamental and main role in an evolutional universe, irreversible, and complex, in which reversibility and simplicity are just particular cases. Man himself comes from time: time, if it was created by man, would be a screen between nature and man himself. This idea of time *prior-to-man* belongs to a vision of continuous evolution of the universe, where gravitation and thermodynamics are in continuous dialectic. According to this point of view, the future of our universe itself is not, up to now, determined, just as human life and society are not. The real message, according to Prigogine, of the second principle of thermodynamics is the impossibility to predict the future, which is open, whether referring to the little systems of physics, or to the totality of the universe of which we are a part. If we observe our universe, we can see that other than mechanical time, there is irreversibility, which implies interior time, chemical time. The difference between a chemical reaction and life is that in the first case, when we stop feeding it, its interior time dies, while for life, interior time continues and flows from one generation to another, from one species to another, becoming more and more complex (Prigogine 1988: 24).

To read the history of our universe as a history of an autonomous time, or of an increasing autonomy of time is one of the interesting temptations of contemporary (Prigogine 1988: 24) science. This temptation has widely interested Prigogine, who asked himself radical and important—if not fundamental—questions, such as: if the universe is mechanical and/or thermodynamic, what was there before? Were there the reversible laws of mechanics, of quantum theory, of relativity or the direction of time? Did Aristotle have a good insight by pointing out the before and the after, or was he wrong? One can find an answer to these problems in Prigogine studies on thermodynamic systems far from equilibrium, cases in which a system, far from being isolated, undergoes strong conditioning from outside (energetic flows or

[1]Quotation translated by the Author.

reactive substances) (Prigogine 1988: 26). And we can understand the world around us just by paying attention to these properties. *Dissipative structures* involve this property of sensitiveness and coherent movements, possibilities of multiple states, and so of historicity of the «choices» adopted by the systems (Prigogine 1988: 26), properties studied by nonlinear physics-mathematics.

If we assume that in conditions of equilibrium each molecule sees what there is around it, while in nonequilibrium conditions, for example, in the case of chemical watches or of the big hydrodynamic flows, it is necessary for there to be some signals go across the entire system, and that material elements can *see* further than their immediate vicinity: matter should become sensible (Prigogine 1988: 26). Life itself is an example of this situation: in fact it should have incorporated physical properties such as gravitation, electromagnetic fields, light, weather, and so on, acquiring the flexibility proper to a substance far from equilibrium.

This flexibility implies many possible properties, many possible states, which are the different dissipative structures accessible (Prigogine 1988: 27). Equations are nonlinear, while they become linear in proximity to equilibrium where there is just one solution: the opposite of the precedent case. But all this implies other difficulties such as the ones relative to attractors,[2] to sensitiveness to the initial conditions, to the *deterministic* case, and so on. Life is the kingdom of the nonlinear, life is the kingdom of the autonomy of time, it is the kingdom of the multiplicity of structures (Prigogine 1988: 28), but all this is better hidden in the un-alive universe, where there are some structures, there is the nonlinear, and thus, the time of evolution becomes longer. Life gives us help to see these things, such as the birth and the death of structures, in a short time, on a reductive temporal scale. But life is time inscribed in substance, as it is for a work of art. It is a symmetry broken as a polymer, as a DNA. But does time have a beginning? How has it appeared in the universe? These are the questions Prigogine poses for himself and this is his singular answer:

time precedes the universe

because the universe itself

[2]In regard to attractors, I must underline that, for example, a pendulum would continue to swing indefinitely if there were no attrition, while movement diminishes to then stop; there is an attractor point which explains this example of asymptote stability. Apart from this simple example of the pendulum we have observed other complex cases in which we can no longer speak of a single attractor point, but of a closed bending that translates a periodic behavior. An attractor point is the result of a set of points to which the system observed is attracted at the beginning by one point, and then by another one, and so on. We are at the presence of a "strange attractor," as this archetype (of chaos) has been called. They can be found in greater or lesser density on some lines, some surfaces, and some volumes. Their dimensions cannot be stated with whole numbers, because they are distributed densely way. Mandelbrot called them "fractals" because they indicate something irregular and indented, for example a coastline (Mandelbrot 1977; Peitgen and Richter 1986; Bellacicco 1980).

> is the result of an instability succeeded to a precedent situation; so the universe would be the result of a phase change on a big choice[3] (Prigogine 1988: 39).

But was there a birth of time? Probably there was a birth of our universe. Is this where the birth of time itself lies? Truthfully

> … already in floating empty space time preexisted at the potential state (Prigogine 1988: 63)

and, Prigogine adds, it is a time which is not our historical, chronological time, it is neither eternity nor the eternal return. Effectively, it is no longer just irreversibility and evolution but is a potential time, a time which is «always already here», in a latent state, which just needs a floating phenomenon to become real. In this meaning time was not born with our universe: time precedes existence, and can make other universes exist (Prigogine 1988: 64). But a definition of time, of this pre-existent time is not yet given us, we do not have the language for it, we are still searching for the language that will clarify this point for us, giving us more and richer words.

Furthermore time, according to Prigogine, has a very important role: a creative one. In each phenomenon we observe we can see the creative role of an irreversible phenomenon, the creative role of time (Prigogine 1988: 79). To the classic conception that considers that irreversibility implies entropy, which on the one hand implies probability, precisely because we do not know the exact trajectories,[4] Prigogine opposes the second principle of thermodynamics as a message regarding the structure of the universe.

He stresses the point that our universe at its beginning was in a state of equilibrium. It is precisely the existence of substance and not of anti-substance that is the proof of the breakage of symmetry (Prigogine 1988: 80), (while in the laboratory we are able to produce the same quantity of substance and anti-substance). And he adds, always keeping a positive reading of the second principle, that

> the evolution of the universe was not in the direction of degradation but in the direction of an increase in complexity, with structures which appear progressively at each level, from the stars and the galaxies to the biological systems (Prigogine 1988: 80).

So time is not an *illusion*. The thesis that our universe shall go towards a decline, due to an exhaustion of resources (this thesis means reading the second principle of thermodynamic in a negative and pessimistic way) does not take into consideration the results found by the studies of *dissipative systems*, those far from equilibrium

[3]Quotations from Prigogine (1988) translated by the Author.

[4]In instable dynamic systems the concept of trajectory has no meaning. In fact "two points, as narrow as you want, will go exponentially far from each other, according to the number called «Ljapuno's exponent». Instability destroys the character of the trajectory and modifies our concept of space-time" (Prigogine 1988: 79). Already Einstein married the concept of time with matter, now we must marry the space-time with irreversibility, that is, "that irreversibility expresses also a structure of space-time" (Prigogine 1988: 79).

and, in more general terms, from complexity. On the other hand *time* is not an illusion in a repetitive horizon of our universe, of cyclical being, because

> the reality of the universe is more complex: on long terms and on a cosmological level gravitation and entropy are implied, and the game between the two is far from being cleared up (Prigogine 1988: 81).

Truthfully, since now whether a vision of decay or of degradation, whether of repetition or endless reproduction seem over-simplified. There is a profound dialectical relation between gravitation and thermodynamics, from the study of which many more rational hypothesis may spring out, which will consider complexity itself. In any case,

> The teaching of the second principle is that this becoming stays open, tied as it is to always new processes of the transformation and increasing of complexity (Prigogine 1988: 81).

In this horizon the role of time remains central and prioritary in its creative, pre-existent or latent aspect, for the sake of those who consider it an illusion and a dissipation. It has allowed our universe *to come to be* not as a unique event, not as a singularity, as a point without any extension in which the totality of energy and of substance of the universe itself is concentrated. And this last hypothesis does not match the relativistic theory, because physical laws are not applicable to this point of infinite intensity of substance and energy.

> It is a process that matches particular conditions[5] (Prigogine and Stengers 1989: 145)

and that could explain the passage from a supposed empty universe to our material world and give an answer to the question "Why is there something rather than nothing?"

So our universe has an age, a direction and a time, where

> irreversibility may not be an added property, which underlines a difference between the effective evolution of our universe and the ideal of an adiabatic evolution, as it's the case of the pattern of inflation. Instead, it may be the essential expression of the genesis of our universe (Prigogine and Stengers: 153)

because

> it may not be energy, but entropy to make the difference between Minkowski's simple space-time empty universe and our material universe (Prigogine and Stengers 1989: 153).

The concept of a universe which goes toward its degradation does not explain its initial highly ordered, and improbable a priori state, while the idea of an irreversible creation of substance explains such a state, because

> the geometrical, space-temporal universe corresponds to a coherent state which is going to be destroyed by entropic creation of substance (Prigogine and Stengers 1989: 153).

---

[5]Quotations from Prigogine and Stengers (1989) translated by the Author.

According to Prigogine, thermic death is at the origin, it is behind us, it is part of the history of our past at the moment when the space-temporal structure of our empty universe broke down, and when, breaking the smooth «space–temporal fabric», substance appeared, and with it, entropy (Prigogine and Stengers 1989: 153). This event for Prigogine corresponds to the instability of an emptiness of a quantum theory, which is the contrary of nothing, containing potentially all possible particles, and so corresponds to an empty original space. All this denies the hypothesis of the birth of universe as a singular event, a hypothesis which implies the creation of spatio-temporal bending. The creatures of the primordial universe must be essentially dissipative, characterized by a very high entropy (Prigogine and Stengers 1989: 156).

The floating vacuum of quantum theory can cause the instability of Minkowski's (Prigogine and Stengers 1989: 156) (empty) universe and the widening of a fluctuation would break Minkowski's space-time, giving birth to our universe by an irreversible production of particles having mass and the bending of space-time (Prigogine and Stengers: 157). This pattern is similar to the processes of nucleation, of the crystallization of a liquid, of its superfusion (with a temperature less than that of crystallization), so this pattern implies that our universe should undergo an inflationary phase, and renders useless the assumption of a cosmological constant so pregnant with difficulty.

The instability which creates Substance replaces the initial singularity, or *big bang*, with its enormous problems, and may explain the passing from an empty universe (of quantum theory) *to a universe in exponential expansion crowded by mini black holes*, with a life lasting 10 s. This is the duration of the birth of our universe, in which there may be produced almost the whole totality of its entropy (Prigogine and Stengers 1989: 151). With the vaporization of black holes, it is possible that the evolution of the adiabatic type described by the standard pattern (Prigogine and Stengers 1989: 158), could begin, while at the death of black holes that generate substance and radiation, the exponential expansion becomes an adiabatic expansion, that of the standard pattern, which still continues nowadays (Prigogine and Stengers 1989: 158). The fundamental conclusion that Prigogine underlines is that the *big bang* is instable in its structure (Prigogine and Stengers 1989: 205–211) and that the project of this pattern matches the description of our actual universe.

The calculations of entropy of the black holes created during the 10 s of the coming to birth of our universe allows, beginning from the value of the three universal constants, a correct evaluation of the data that characterize the thermodynamic structure of our actual universe: the entropy (measured with the number of the photons of the universe) and, mostly, the relation between the number of particles with a mass, and that of the photons which constitute this universe (Prigogine and Stengers 1989: 159). So why is the characteristic of the birth of substance irreversible, beginning from space-time? In the Prigoginian thesis, the creation of our universe is followed by a decrease of gravitational energy. If we assume that the vacuum of quantum theory is an energetic and basic state of non-value, the creation of our material universe then gives a negative value to that state, which corresponds to the bending of space-time that is given as a consequence (Prigogine and Stengers 1989: 159). This implies that our universe is in expansion, because the transferring of energy is in one direction (gravitational energy towards material energy), while the opposite is impossible. I want to stress the point here that the expansion and creation of substance imply each other, but the period of the expansion of creation of our universe is distinguished from the expansion without creation: see the cosmological equations of Friedmann-Lemaitré.

This pattern describes the birth of our universe in contradiction to other patterns, and implies that the time of our universe is a different thing from Time and its birth.

According to Prigogine *the original vacuum of quantum theory* implies a latent irreversible time, presupposed by this floating vacuum (Prigogine and Stengers 1989: 160). In fact, the conditions of nonequilibrium do not create the arrow of time, but allow its manifestation at a macroscopic level, so that the arrow of time of our universe is not created, but actualized by the fluctuation that gives birth to this universe. That is, *time precedes existence*.

According to this point of view, we may exclude the birth of our universe as a singularity and thus as a unique event, but we may not exclude the above-mentioned conditions of nonequilibrium from re-proposing themselves. This may imply an end of our universe, but Prigogine affirms that now there is no reason to make such a hypothesis; thermic death is our past, but we cannot exclude a reproduction of the initial conditions of *the floating vacuum of quantum theory,* nor can we exclude a *death of our universe* which could lead to *a new floating empty vacuum*, from which, once again having the conditions of nonequilibrium, a new universe would arise. Finally, our *universe is a possible one*, among *many possible universes*. But *have* these possible universes *already been* or *must they come to be* in respect to *latent time* or to our *time* or to *time* of the possible? Of course they must be in respect to latent time. But since it is a different possibility with respect the realized one, the actualized one, the measure of time cannot be established with respect to possible time, because we cannot grasp it in any way, we know nothing of a possible universe different or similar to ours. The only relation we can find is that with our time, but this is not the original one because it is just one particular and precise actualization of it. It is the time of the birth of our universe and cannot be considered as a reference.

This recalls Miletus' *physiology*, according to which the becoming of our universe has an origin that permeates time; but we must ask, then, if time could also work as an organizer as it seems, and so if time could be identified with Heraclitus' *logos* in addition to Miletus' element.

But the thesis of change—because it is change—of the *possible universes*, as of *our own universe*, needs so a strong postulate: it is enough to think of a change from *a latent state*, which pretends to be more than a simple potentiality, because it is a *floating vacuum*, to another state of evolution 10-, which correspond to the duration of the birth of our universe, up to the exponential growth of entropy, followed by the adiabatic phase (with constant entropy)? So we should speak of duration, of change instead of absolute time, that should precede every existence and every thought, and this would be very near to what we call *eternity*. In change, in duration, *states of local conflict* are possible and tolerable, as it seems to emerge from Prigogine's pattern, which is of course much more reliable than others for its unitary and problem-solving vision, which other patterns lack.[6] We must not stop at the *thresholds* of *entropic explosions* and presuppose absolute time, which does not match the capabilities of our language and our imagination (Prigogine and Stengers 1989: 163). Should we then give up thinking of its origin, the origin of *absolute time or eternity*? This presupposes the time of our universe, or better said, the times of our universe, as they were captured by different and various correlative approaches. To take *change* instead of a *temporal horizon*, inherent in that floating vacuum of quantum theory, the thresholds where to stop seem to fall and change seems to bring things totally into harmony with Prigogine's pattern that we have analyzed before.

In this horizon, we must accept what Stephen Hawking (Hawking 1988)[7] himself said about the beginning and the end of time, in physic-cosmological meaning, as we can accept the meaning of experienced time, which stands in the evolution of our history, which begins with us, coincides with the origin of our biological time, or of our biological times, to end with the end of our biological history on a macroscopic scale. Our evolutionary history is, of course, underlined by experienced time, the charioteer of our changes, but in a dialectical relation with chronological, chronometric, chronosophic times, in a relation of one among many, which produces states of suffering. But it does not make this evolutionary history less interesting. So, at the end of our journey, at least the consciousness of being "inhabitants" of time and *bearers* of change appears, and this occurs whether we have behind us *big cataclysms* (Eliade 1963) or thermic Death (Prigogine 1988).

If our future really has a goal (of salvation or of ruin or in any case, a goal) we shall meet it as the haunters of *residences* in which "there are not marble idols/not metal enticements/but rough wood/ready to rot/with its shapes" (Rigon 1983). So we may not suffer change anymore, which will not be the old, the painfully lost, that we try to restore, disguising and selling it as new, but what springs out (Masullo 1993)[8] and which implies either human contingency (Monod 1970: 45–46) or external and interior irreversibility.

---

[6]For alternative patterns see Lerner (1992).

[7]Hawking (1982); see also Hawking (1988).

[8]In this work Masullo carries out a severe philosophical analysis of the Prigoginian concept of irreversibility, suitably distinguishing the evolutionary irreversibility of Prigogine from the one of the foregoing thermodynamics.

References

Bellacicco, Antonio. (1980). *La rappresentazione frattale degli eventi.* Firenze: La Nuova Italia Scientifica.

Eliade, M. (1963). *Aspects du Mythe.* Paris: Gallimard.

Hawking, S. (1982). Ai margini dello spaziotempo. In P. Davies (Ed.). *La nuova fisica*, (pp. 66–74) (D. Canarutto, Trans.). Torino: Boringhieri.

Hawking, S. (1988). *A Brief history of time.* United States and Canada: Bantam Books.

Lerner, E. (1992). *The big bang never happened.* New York: Viking Press.

Mandelbrot, B. (1977). *Fractals: Form, chance and dimension.* San Francisco: W. H. Freeman.

Masullo, A. (1993). Epistemologia dell'irreversibile ed etica del tempo. *Paradigmi, 11*(31), 125–152.

Monod, J. (1970). *Le hasard et la nécessité.* Le Seuil: Essai sur la philosophie naturelle de la biologie moderne. Paris.

Peitgen, H., & Richter, P. (1986). *The beauty of fractals.* New York: Springer-Verlag.

Prigogine, I. (1988). *La nascita del tempo.* (B. Pedretti, Trans.). Roma-Napoli: Theoria.

Prigogine, I., & Stengers, I. (1989). *Tra il tempo e l'eternità.* (C. Tatasciore, Trans.). Torino: Bollati Boringhieri.

Rigon, F. (1983). *Dimore.* Torino: Einaudi.

Von Wright, & Henrik, Georg. (1974). Tempo cambiamento e contraddizione. In Claudio Pizzi (Ed.), *Logica del tempo.* Torino: Boringhieri.

The End of Time: New Perspectives of Self-identification for Man

Gianluca Giannini

Time is waste of money
Oscar Wilde

Abstract At any angle it is constituted in our Tradition, or in addition to *chronos*, as *aion*, *kairos,* and *eniautos*, the concept of time has been (and is) the fundamental reason of our self-identification, self-comprehension, and self-narrating. This paper, through the reconstruction of some of milestones of Western Philosophy until post-Einstein physics, tries to analyze Julian Barbour's proposal. He argues that the holy grail of physicists—the unification of Einstein's general relativity with quantum mechanics—may well spell the end of time. The idea of the discontinuity of time proposed by Barbour attempts to explain in a theoretical context a universe composed of many points he calls 'Now'.

At any angle it is constituted in our Tradition, or in addition to *chronos*, as *aion*, *kairos,* and *eniautos*, the concept of time has been (and is) the fundamental reason of our self-identification, self-comprehension, and self-narrating. As well as—together with space—the fundamental way of our understanding of the world.

But something, something relevant, happened in the Twentieth Century: *time is dead*.
What does that mean?
Does it mean that Western Man is dead?
Does it mean that he is no longer able to indentify himself, narrate himself, and understand the world?
What happened?

In his *Out of Time's Joints*, especially in the Kantian Lesson of March 14, 1978, dedicated to *Synthesis and Time*, Gilles Deleuze observed that:

> Time cannot be defined by succession because succession is only a mode of time, coexistence is itself another mode of time. You can see that he [Kant] arranged things to make the simple distribution: space-coexistence, and time-succession. Time, he tells us, has three

G. Giannini (✉)
Philosophy Section, Department of Humanities, University of Naples Federico II, via Porta di Massa 1, 80133 Naples, Italy
e-mail: gianning@unina.it

F. Santoianni (ed.), *The Concept of Time in Early Twentieth-Century Philosophy*, Studies in Applied Philosophy, Epistemology and Rational Ethics 24,
DOI 10.1007/978-3-319-24895-0_29

> modes: duration or permanence, coexistence and succession. But time cannot be defined by any of the three because you cannot define a thing through its modes (Deleuze 1978).

Beyond his deep analysis, Deleuze tell us that we cannot define ourselves as *ànthropos* through duration or permanence, coexistence, and succession. We cannot define ourselves through Time.

Why does Deleuze, through Kant, highlight this limit?

Because he is aware that Einstein's revolution has undermined not just modes of a measurement unit, but *the* chance of self-identification and self-comprehension by *ànthropos*.

Simply, time is out of joint. Time has evaporated.

In order to understand the direction of the turn in the Twentieth Century, the meaning of '*time is dead*', we must grasp the reasons of time's fixation in the story of Western Man.

So we must consider that in his *Timaeus*—that gives us a mythical account of the making of the world by a Demiurge, who seeks to create the universe by imposing form on chaotic matter—Plato, concerning the severe question of time, wrote:

> The father […] resolved to have a moving image of eternity, and when he set in order the heaven, he made this image eternal but moving according to number, while eternity itself rests in unity; and this image we call time. For there were no days and nights and months and years before the heaven was created, but when he constructed the heaven he created them also. They are all parts of time, and the past and future are created species of time, which we unconsciously but wrongly transfer to the eternal essence; for we say that he 'was,' he 'is,' he 'will be,' but the truth is that 'is' alone is properly attributed to him, and that 'was' and 'will be' are only to be spoken of becoming in time, for they are motions, but that which is immovably the same cannot become older or younger by time, nor ever did or has become, or hereafter will be, older or younger, nor is subject at all to any of those states which affect moving and sensible things and of which generation is the cause. These are the forms of time, which imitates eternity and revolves according to a law of number. Moreover, when we say that what has become *is* become and what becomes *is* becoming, and that what will become *is* about to become and that the non-existent *is* non-existent—all these are inaccurate modes of expression (Plato 1871: 37d–38b).

If Time is *a moving image of eternity*, this is not to say that Time is change and movement.

That Time is not change and movement—because they are always referred to something that moves and changes—is then fixed permanently by Aristotle. In the tenth chapter of the fourth book of *Physics* he found that:

> As time is most usually supposed to be motion and a kind of change, we must consider this view. Now the change or movement of each thing is only in the thing which changes or where the thing itself which moves or changes may chance to be. But time is present equally everywhere and with all things. Again, change is always faster or slower, whereas time is not: for 'fast' and 'slow' are defined by time—'fast' is what moves much in a short time, 'slow' what moves little in a long time; but time is not defined by time, by being either a certain amount or a certain kind of it. Clearly then it is not movement. (We need not distinguish at present between 'movement' and 'change') (Aristotle 1930: IV 10, 218b 9–19).

Neither does Time exist without change and movement. So the Stagirite could have considered:

> For when the state of our own mind does not change at all, or we have not noticed its changing, we do not realize that time has elapsed, any more than those who are fabled to sleep among the heroes in Sardinia do when they are awakened; for they connect the earlier 'now' with the later and make them one, cutting out the interval because of their failure to notice it. So, just as, if the 'now' were not different but one and the same, there would not have been time, so too when its difference escapes our notice the interval does not seem to be time. If, then, the non-realization of the existence of time happens to us when we do not distinguish any change, but the soul seems to stay in one invisible state, and when we perceive and distinguish we say time has elapsed, evidently time is not independent of movement and change. It is evident, then, that time is neither movement nor independent of movement (Aristotle 1930: IV 11, 218b 22–35 219a 1–2).

We perceive movement (and change) and time together: we think, through the intellect (*noûs*), time. The time that has passed is always thought to be in proportion to the movement (or change). Time is a property of movement (and/or change): only from this explanation does there occurs the distinction of 'before' and 'after'.

Time is number of motion in respect to 'before' and 'after': *it is a kind of number*. "The 'now' measures time, insofar as time involves the 'before and after'" (Aristotle 1930: IV 11, 219b 12). So, "Time is a measure of motion and of being moved, and measures the motion by determining a motion which will measure exactly the whole motion" (Aristotle 1930: IV 12, 221a1–3).

It is a continuous measure, a continuous proportion, and then *to be in time* means that things (and us), their (our) being should be measured by time.

Western Tradition, over several centuries, in some way dominated by circular conception of time since the extreme meaning of the Eternal Return, has registered the turning point of linear time introduced by Christianity.

In his *Confessions* Saint Augustine introduced this concept in the following way:

> What then is time? If no one asks me, I know: if I wish to explain it to one that asketh, I know not: yet I say boldly that I know, that if nothing passed away, time past were not; and if nothing were coming, a time to come were not; and if nothing were, time present were not (Agustine 1909: B. XI, chap. 14).

We only find certainty in Creation: "At no time then hadst Thou not made anything, because time itself Thou madest" (ibid.). So, only in this way can we finally say:

> What now is clear and plain is, that neither things to come nor past are. Nor is it properly said, 'there be three times, past, present, and to come': yet perchance it might be properly said, 'there be three times; a present of things past, a present of things present, and a present of things future'. For these three do exist in some sort, in the soul, but other where do I not see them; present of things past, memory; present of things present, sight; present of things future, expectation. If thus we be permitted to speak, I see three times, and I confess there are three. Let it be said too, 'three be three times, past, present, and to come': in our incorrect way (Agustine 1909: B. XI, chap. 20)

Finally, it is from this theoretical articulation that Western Man has discovered the endless possibilities related to time.

Time as measurement and computational unit in the terms of past, present, future.
Time as measurement and computational entity in the linear terms of before, now, and after.
Time as self-measurement and self-computation/comprehension entity in the existential terms of memory (past), sight (present), expectation (future).

Philosophy, for centuries, has been moving in this groove. Western Man, for centuries, has been moving in this groove. Time—with Isaac Newton—has become *absolute*, like an invisible river that flows uniformly forever. In the *Scolium* of *Definition VIII* in *Mathematical Principles*, he wrote:

> I. Absolute, true, and mathematical time, of itself, and from its own nature flows equably without regard to anything external, and by another name is called duration: relative, apparent, and common time, is some sensible and external (whether accurate or unequable) measure of duration by the means of motion, which is commonly used instead of true time; such as an hour, a day, a month, a year. [...]
> Absolute time, in astronomy, is distinguished from relative, by the equation or correction of the vulgar time. For the natural days are truly unequal, though they are commonly considered as equal, and used for a measure of time; astronomers correct this inequality for their more accurate deducing of the celestial motions. It may be that there is no such thing as an equable motion, whereby time may be accurately measured. All motions may be accelerated and retarded, but the true, or equable, progress of absolute time is liable to no change. The duration or perseverance of the existence of things remains the same, whether the motions are swift or slow, or none at all: and therefore it ought to be distinguished from what are only sensible measures thereof; and out of which we collect it, by means of the astronomical equation. The necessity of which equation, for determining the times of a phenomenon, is evinced as well from the experiments of the pendulum clock, as by eclipses of the satellites of Jupiter. [...] As the order of the parts of time is immutable, so also is the order of the parts of space. Suppose those parts to be moved out of their places, and they will be moved (if the expression may be allowed) out of themselves. For times and spaces are, as it were, the places as well of themselves as of all other things. All the things are placed in time as to order of succession; and in space as to order of situation. It is from their essence or nature that they are places; and that the primary places of things should be moveable, is absurd. These are therefore the absolute places; and translations out of those places, are the only absolute motions (Newton 1846: 74–75).

Leibniz, for instance, not necessarily in conflict but, in a way, in continuity with this argument, argues that the notion of time is simply the way in which we, as finite intellects, perceive what are essentially a series of intelligible relations between things. Time is simply 'well-founded phenomenon' by which we inadequately perceive the true conceptual order of things—ourselves included.

On this general and inclusive line, in Kant's proposal, time concerned with the conditions under which a subject can attain, a priori, knowledge of a world of objects.

Into a long passage of the first *Critique*, in fact we can read:

> Time is the formal condition a priori of all phenomena whatsoever. Space, as the pure form of external intuition, is limited as a condition a priori to external phenomena alone. On the other hand, because all representations, whether they have or have not external things for their objects, still in themselves, as determinations of the mind, belong to our internal state; and because this internal state is subject to the formal condition of the internal intuition, that is, to time—time is a condition a priori of all phenomena whatsoever—the immediate condition of all internal, and thereby the mediate condition of all external phenomena. If I can say a priori, 'All outward phenomena are in space, and determined a priori according to the relations of space', I can also, from the principle of the internal sense, affirm universally, 'All phenomena in general, that is, all objects of the senses, are in time and stand necessarily in relations of time' (Kant 1855: A34/B50–51).

And then:

> What we have now set forth teaches, therefore, the empirical reality of time; that is, its objective validity in reference to all objects which can ever be presented to our senses. And as our intuition is always sensuous, no object ever can be presented to us in experience, which does not come under the conditions of time. On the other hand, we deny to time all claim to absolute reality; that is, we deny that it, without having regard to the form of our sensuous intuition, absolutely inheres in things as a condition or property. Such properties as belong to objects as things in themselves never can be presented to us through the *medium* of the senses. Herein consists, therefore, the transcendental ideality of time, according to which, if we abstract the subjective conditions of sensuous intuition, it is nothing, and cannot be reckoned as subsisting or inhering in objects as things in themselves, independently of its relation to our intuition. This ideality, like that of space, is not to be proved or illustrated by fallacious analogies with sensations, for this reason—that in such arguments or illustrations, we make the presupposition that the phenomenon, in which such and such predicates inhere, has objective reality, while in this case we can only find such an objective reality as is itself empirical, that is, regards the object as a mere phenomenon (Kant 1855: A35-36/B52–53).

In order to these philosophical views, we tend to believe that destiny is not fixed and that all time past fades into oblivion. However, we are *in* time. We *are* time. Time is what defines us, what allows us, at each instant, to place ourselves, understand the outside world and build realities.

The passing time not only describes, but gives meaning and significance to the becoming of things. It gives meaning and significance to '*ànthropos*' life and death.

But, as already mentioned, in the twentieth century, something shocking happened.

Albert Einstein had demonstrated that temporal reality is relative to each object in the universe, and that time is a 'subject' inseparable from space. Even specialists who synchronize time in the world are aware that the world is handled by an arbitrarily stipulated ticking, as clocks are unable to measure time at all. Apparently, the only alternative is to sink into a 'temporary illusion' of this infinity, knowing that there is a space where our past still exists and what we do does not change. According to Einstein's *Theory of Relativity*, measurements of various quantities are relative to the velocities of observers: in particular, we cannot speak of simultaneity. We must speak of *relativity of simultaneity*: two events, simultaneous for one observer, may not be simultaneous for another observer if the observers are in motion. We can say that there is no such thing as absolute space. There is no such thing as absolute time. Newton's foundation for all of physics was flawed.

By rejecting absolute time, Einstein rejected the notion that everyone, regardless of his or her motion, must experience the flow of time in the same manner. Time is relative, Einstein asserted. Each person travelling in his or her own way must experience a different time flow than others, traveling differently. What I call space must be a mixture of your space and your time, and what you call space must be a mixture of my space and my time. You might be tempted to assert that this 'mixing of space and time' is nothing but a complicated way of saying that 'simultaneity depends on one's state of motion' (Thorne 1994).

Henceforth, space by itself, and time by itself, are doomed to fade away into mere shadows, and only a kind of union of the two will preserve an independent reality. But time, as we have long conceived it, simply no longer exists.

As Einstein himself would say in one of his last *Letters*: "People like us, who believe in physics, know that the distinction between past, present, and future is only a stubbornly persistent illusion."

Therefore, time does not exist?

It's unreal?

Simply put: *time is dead*?

Saying that time is unreal could mean denying so-called 'temporal becoming': the idea is that past and future things, events and states of affairs (or however one conceives the material contents of space-time) are just as 'real' as present ones. On the other hand, saying that time is unreal could mean the opposite: for instance, *presentism*. The idea is that only the present is real: past and future things are unreal. As regards the main idea of *presentism*, it does not matter how you conceive the material contents of space-time, though of course in more precise versions, it can matter. Thus, the debates about these two positions turn on the contrast between the real and the unreal.

But there is, now, a third conceptual form.

An interpretative form, capable of dissolving the reasons of the debate between the real and the unreal.

We can speak of *spontaneity*.

What does it mean?

Spontaneity presupposes the idea of a set of many possible courses of history, where each course of history is a 'block universe.' But *Spontaneity* then proposes that unbeknownst to us, the actual history jumps between disparate instantaneous states.

From the development of this new form, the British physicist, Julian Barbour, author of *The End of Time: The Next Revolution in Physics*, explains that in a special dimension, time simply doesn't exist.

"If you try to get your hands on time, it's always slipping through your fingers. People are sure that it's there but they can't get hold of it" (Barbour 2001).

Barbour argues that the holy grail of physicists—the unification of Einstein's general relativity with quantum mechanics—may well spell the end of time. The concept of a timeless universe is not only irresistibly attractive to a handful of scientists, but such a model may pave the way to explain many of the paradoxes that modern physics faces in explaining the universe. We tend to think and perceive time to be linear in nature, the course of which inevitably flows from past to future. This is not only a personal perception of all humans, but also the context in which classical mechanics analyzes all mathematical functions within the universe. Without such a concept, ideas such as the principle of causality and our inability to be present simultaneously in two events would begin to be addressed from a completely different level. The idea of the discontinuity of time proposed by Barbour attempts to explain in a theoretical context a universe composed of many points he calls 'Now.' But such 'Nows' would not be understood as fleeting moments that came from the past and will die in the future; a 'Now' would only be one among the millions now existing in the eternal universal mosaic of a special dimension impossible to detect, each one related in a subtle way to the others, but none more outstanding than the neighboring one. They all exist at the same time. With such a mix of simplicity and complexity, Barbour's idea promises a great relief to anyone who is willing to accept the lack of time before the Big Bang.

He remarks:

> In normal physics, with a notion of time, Zeno's paradox is readily resolved. However, in my timeless view the paradox is resurrected, but the arrow never reaches the target for a more basic reason: the arrow in the bow is not the arrow in the target (Barbour 1999: 49).

Here we find Heraclitus' image of the river: what we perceive as a river is never the same twice; nothing moves because there is no time sequence; nothing stays the same for two periods; and our sense of continuity and persistence is a confusion about the new creation. In this world,

> All around *Now*, along the filament and to either side of it, are other *Nows* with slightly different versions of yourself. All such *Nows* are 'other worlds' in which there exist somewhat different but still recognizable versions of yourself. In other filaments are worlds you would not recognize at all (Barbour 1999: 55–56).

We have, then, a configuration space, which Barbour calls *Platonia*, where

> time truly does not exist. This also applies to motion: the suggestion is that it too is pure illusion. If we could see the universe as it is, we should see that it is static. Nothing moves, nothing changes (Barbour 1999: 39).

In this world, "All we know is that the present Now is real" (Barbour 1999: 53).

At this point of the argument, it is probably inevitable to ask: if infinitesimal fractions of 'Nows' are not connected to each other, how do I remember, for example, what I ate for dinner? Or: if the future is already there, why strive at all?

Such dilemmas have arisen from the illusory perception that time is fleeting, like water in a river. Conversely, the endless 'Nows' independent of each other would not be dispersed. They still make up a structure. They are a block. And this is Barbour's theory: in a space of the cosmos, the future (our future) is already there, deployed, and every second of our past is also present, not as a memory but as a living present.

This means '*living without time*' (Barbour 1999).

Ultimately:

> My basic idea is that time as such as does not exist. There is no invisible river of time. But there are things that you could call instants of time [...]. As we live, we seem to move through a succession of Nows, and the question is, *what are they*? They are arrangements of everything in the universe relative to each other in any moment [...]
> [...] Simply keep this idea of many different things coexisting at once in a definite mutual relationship. The interconnected totality becomes my basic thing, a Now. There many such Nows, all different from each other. That's my ontology of the universe – there are Nows, nothing more, nothing less (Barbour 2001).

References

Aristotle. (1930). *Physics.* (R.P. Hardie & R.K. Gaye, Trans.). Oxford: Clarendon Press.
Augustine. (1909). *Confessions.* (E. Bouverie, Trans.). London: Pusey.
Barbour, J. (1999). *The end of time: The next revolution in Physics.* Oxford New York: Oxford University Press.
Barbour, J. (2001). *The end of time: a talk with julian barbour.* In *KurzweilAI.net* 3. Retrieved from http://www.kurzweilai.net/the-end-of-time-a-talk-with-julian-barbour.
Deleuze, G. (1978). *Time and synthesis.* (M. McMahon, Trans.). Retrieved from htpp://www.webdeleuze.com/php/texte.php?cle=66&groupe=Kant&langue=2.
Isaac, N. (1846). *The mathematical principles of natural philosophy.* Adee: Trans A. Motte New York.
Kant, I. (1855). *The critique of pure reason.* (J.M.D. Meiklejohn, Trans.). London: Bohn.
Plato, (1871). *Timaeus* (p. 1871). Trans. J. Jowett. Oxford: Clarendon Press.
Thorne, S. K. (1994). *Black holes and time warps. Einstein's outrageous legacy.* New York: Norton & Company.

Commentary: Gustave Guillaume Between Linguistics and Philosophy of Language: A New Point of View

Louis Begioni

In this contribution, Rocco Pititto presents a reflection on the linguistics of Gustave Guillaume as natural starting point Ferdinand de Saussure, the father of general linguistics. His comment emphasizes the importance of the theoretical approach of Gustave Guillaume that was too often sidelined and almost ignored by the major currents of the twentieth century linguistics. He shows the commitment of Gustave Guillaume to studies that take into account both diachronic and synchronic aspects.

But according to Pititto, the originality of Gustave Guillaume is to have inserted the language into the operations of thought and memory which enables him to offer a vision of Gustave Guillaume as part of the philosophy of language and mind; the description of language and the study of meaning refer to two different types of knowledge, to linguistics on the one hand and to philosophy on the other, in particular, the philosophy of mind.

The paradigms of Saussurian linguistic concepts could also be referred to the most diverse fields of knowledge, from anthropology to architecture, from philosophy to the social sciences, from literature to psychoanalysis and from morality to meaningful practices. The birth of semiology, first, and semiotics, later, represented the triumph of the linguistics of Saussure, because it made possible the transformation of linguistics itself into a general conception of culture, as if it were a new metaphysics in the context of the society of the twentieth century, which had even rejected the return of all forms of metaphysics. The crisis of the post-Saussurian linguistics of the twentieth century is at the origin of today's new interest in the linguistic conceptions of Guillaume.

Among the theoretical concepts, we have to stress the central dichotomy *langue/discours* that has not exactly the same meaning of Saussure's *langue/parole*. The *discours* is the temporary use of the *langue* by a speaker. It is the sum of systematized means that human thought has internalized. It is not comparable to the word *parole* in the Saussurian meaning. For Guillaume, *parole* exists both in the *langue* (phonological analysis) and in the *discours* (phonetic analysis). In the act of

L. Begioni (✉)
University Charles-de-Gaulle–Lille 3, Langues Étrangères Appliquées,
14 Place Bodart Timal, 59058 Roubaix, France
e-mail: louis.begioni@uniroma2.it

F. Santoianni (ed.), *The Concept of Time in Early Twentieth-Century Philosophy*,
Studies in Applied Philosophy, Epistemology and Rational Ethics 24,
DOI 10.1007/978-3-319-24895-0_30

language, Guillaume identifies the level of power (the *langue*) and the level of the act (the *discours*). Both levels are structured under the sign of temporality, meant by Guillaume as an interpretative paradigm of the linguistic act. Language as a unit of *langue* and of *discours* is not the sum of different faculties, but a process that is divided into two distinct phases of *langue* and *discours* and is characterized as a continuous passage from the level of *langue* to that of *discours*. This step, however, would not be conceivable without a temporal support. In the linguistic phenomenon, there is an accomplished task that is actualized in an infinitesimal duration. It is an infinitesimal temporality which Guillaume named *operative time* (*temps opératif*). It is the central concept of his approach to the problems of language, the reciprocity of the relationship between language and discourse and the idea of succession in the process of the construction of language. Gustave Guillaume's merit is to consider human language in the temporal dimension of the operations of thought, making it an interpretative paradigm of language itself. It is a paradigm able to give a sufficiently well-founded explanation to a number of linguistic phenomena, without which they would not find a sufficient explanation. According to him, thought is the place of definition of time, but time is the site of action of thought. Being ahead of his time, he was taken into consideration by the majority of linguists.

Today the psychomechanics of language can open up new research perspectives giving interesting answers when cognitive linguistics seems to have already exhausted most of its capabilities. For this reason, some linguists, followers of Guillaume, are trying to find a new methodological approach backing up on renewed theoretical concepts.

Commentary: Einstein, Prigogine, Barbour, and Their Philosophical Refractions

Lorenzo Magnani

In the chapters of Principe, Grana, and Giannini, the epistemological core is rooted in various aspects inherent the concept of time which rethink some classical interpretations of it—such as the relativity, the dissipative structures of dynamic systems, and the idea of time as a continuum. All these aspects develop intertwined refractions in the philosophical entanglement.

Salvatore Principe exposes accurately the epistemological innovations introduced by Einstein's theory of relativity regarding the concept of time. The ideas of an absolute length and space, of a time and a stream of absolute time, are metaphysical concepts that go beyond what the experience and the observer can justify. For classical mechanics, our everyday mathematical and physical notion is based on the idea that a timing interval and a space interval between two phenomena are always the same for any observer, in all conditions of observation. Through the critical examination of the concept of simultaneity, by taking up the arguments of Einstein, Principe showcases how the measures of time intervals and lengths of space made by inertial observers do not necessarily correspond, giving rise to phenomena like the dilation of time and the contraction of length.

A relativistic interpretation of these phenomena has allowed Einstein to gather three-dimensional space and time into a single four-dimensional entity called *chronotope*, or space–time. Stressing how we can see in Einstein's theory that isolated systems are only an abstraction (or special cases) in nature, and instead open systems—exchanging energy with neighboring systems and thus constantly changing—are the rule, Principe leads us to understand how the complex reasoning concerning dissipative systems was introduced by an initial questioning of classical physics/metaphysics.The relativizing of the systems of reference and the entrance into crisis of the concept of invariance were the opening words of quantum physics and the theory of complexity.

L. Magnani (✉)
Philosophy Section, Department of Humanities, University of Pavia,
Piazza Botta, 27100 Pavia, Italy
e-mail: lmagnani@unipv.it

F. Santoianni (ed.), *The Concept of Time in Early Twentieth-Century Philosophy*,
Studies in Applied Philosophy, Epistemology and Rational Ethics 24,
DOI 10.1007/978-3-319-24895-0_31

Nicola Grana analyzes the concept of time in Prigogine, keeping as theoretical reference the main idea supporting the work of the Brussels school, according to which irreversibility is closely linked to the notion of dynamic instability. In predicting the behavior of instable systems, in fact, it is not our lack of knowledge that is at play, but rather the dynamic nature of the system. Therefore, it is dynamic instability that is at the origin of the notion of probability, and not vice versa. To clarify the meaning of this affirmation, it is sufficient to recall how, for Prigogine, subjecting a particular kind of system to a given constraint, we may obtain as a result an increase in entropy correlated, at the same time, to the delineation of a phenomenon of order.

The underlying mechanism of this kind of phenomenon is, essentially, an amplifying mechanism of the fluctuations. Far from equilibrium, there occurs an amplification of the fluctuations that open the way to a series of varied possibilities. The thermodynamics of nonequilibrium specifically handles systems that exhibit exchanges with the environment, systems in which the variation of entropy is linked not only to processes that occur within the system, but also to flows of energy and matter between the system and the environment. In this kind of system, the decisive physical quantity is not entropy, but the production of entropy, the variation of entropy by unit of time in relation to the processes that occur inside the system. In such systems, therefore, in contrast with thermodynamic systems in equilibrium, in which the equilibrium is linked to the fall toward the most probable or least ordered state, the flow of matter and energy constitutes a driving force that generates order.

In this perspective, Grana carries out his analysis by illustrating the cosmological model, proposed by Prigogine, of a universe that demonstrates at the same time an age (an origin) and an arrow of time. According to this viewpoint, the symmetry of the relations that Einsteinian cosmology established between space–time and matter, inherited from the Newtonian theory of masses in gravitational interaction, is broken: matter is distinguished by space–time by the fact that it is a bearer of the entropy of the universe. Its existence is no longer a given, as the standard model presupposes, but is rather the product of an irreversible process of creation. The initial singularity linked to the Big Bang is substituted thus by the instability of a primordial empty universe in which space–time would be curved, radiating matter. The meaning of irreversibility thus undergoes a radical change, since irreversibility should no longer be linked to an evolution that leads inexorably toward an inert state of the universe (thermic death), but to its birth or perhaps to an eternal succession of universes that are born everywhere and that head toward the infinite.

As Grana correctly points out, for Prigogine time precedes existence, since the conception of the original quantum void he envisages implies, per se, a latent irreversible time presupposed by the fluctuations of this void. This is similar to the situation regarding the states of equilibrium where the conditions of nonequilibrium do not create the arrow of time of our universe, but allow it (always present in the dynamic of the post collisional correlations) to manifest itself at the macroscopic level. In the same way, the arrow of time of our universe is not created, but actualized by the fluctuations that favor its deployment. In other words, for Prigogine it is not possible to think of the origin of time, but only of the "entropic explosions" that presuppose it and which are the creators of new temporalities. In this sense, the dualism between repetition (invariance) and disintegration (dissolution) is overcome by the concept of time understood as construction, a construction that appears to our eyes, simultaneously, as creation and as rediscovery, even if this same creation passes through specific states of invariance and degradation.

Gianluca Giannini confronts the topic of the destructuring of the *realist* conception of time. The end of time, the title of his contribution, alludes not only to the end of linear time (though this is not the only possible form of time) but to the end of temporality conceived as the *virtuality* of a past that is no more and of a future that is not yet. The existential, as well as the ontological and physical condition of time, defines our identity as subjects and as species.

Giannini takes his point of departure from Deleuze's consideration in *Time and Synthesis* and from the link with Kant. Deleuze, in particular, perceives that time cannot be defined. Every attempt in this direction defines the object through its attributes, but does not manage to clarify its nature. A real definition of time appears even more difficult when—after the Einsteinian revolution—time seems to have "evaporated," and with it, the possibilities to define and determine human identity. He continues to carry out his analysis by examining the Platonic conception of time expounded in the *Cratylus* and the Aristotelian treatment of this theme in the *Physics*. Time exists even without its changes, yet it is only through them that it can be thought. The absolute conception of time found in Newton owes to the Platonic and Aristotelian idea. Likewise, as Giannini correctly notes, the correlation between time in itself and the possibility to think of it through movement is found in Augustine. In the Christian and linear conception of time, the time of existence, lived time, is not in contradiction with absolute time; on the contrary, it depends upon the linearity and eschatology of the latter. All this is put into question by Einstein's theories. In particular, the past, present, and future are found to be lacking in any distinction between them. Time is described by simultaneous and equally present instants, subsisting in space.

In Einstein's view, and in Barbour's text *The End of Time*, time is a continuum, which foresees the simultaneity of events and positions. Despite the fact that Bergson's critique of Einstein focuses precisely on the relationship of time with space—reading these authors through Barbour's conception of time as a homogeneous continuum—one almost has the desire to bring the two authors closer together.

Commentary: Reichenbach's Verbal Tenses in the Context of Discovery About Computing Systems

Guglielmo Tamburrini

Abstract This contribution analyzes present applications of temporal logics that are meaningfully related to Hans Reichenbach's groundbreaking work on verbal tenses and their underlying logical structure. Specifically, some formal methods in theoretical computer science will be discussed that enable one to advance empirical hypotheses and to make predictions about the temporal evolution of computing system's behaviors.

Michele Malatesta emphasizes the groundbreaking character of Hans Reichenbach's reflections on conversational language. In particular, he brings out the novelty and rigor of Reichenbach's analysis of verbal tenses—which builds on the distinction between points of speech, event, and reference—and explores its significance for later work on temporal logics. Here, I will briefly point to present applications of temporal logics that are meaningfully related to those early developments in the analysis of verbal tenses and their underlying logical structure. Specifically, I will selectively discuss some formal methods in theoretical computer science that enable one to advance empirical hypotheses and to make predictions about the temporal evolution of computing systems behaviors.

Temporal evolutions in the behavior of computing systems raise challenging prediction problems. Occasionally, disastrous consequences flowed from the inability to predict these behaviors of computing systems. For example, the failure of the software system controlling the radiation therapy machine Therac-25 exposed cancer patients to radiation overdoses and caused the death of six of them in the 1980s. The crash of an Ariane-5 missile shortly after launch in 1996 was similarly caused by a defect in control software. In both accidents, violations of crucial requirements occurred, which concerned the proper behavior of those computing systems. In computer science, requirements on the behavior of computing systems are usually referred to as *specifications*. In their turn, the latter are formulated as properties that runs (or executions) of computing systems are expected to satisfy.

G. Tamburrini (✉)
Department of Electrical Engineering and Information Technologies,
University of Naples Federico II, via Claudio 21, 80125 Naples, Italy
e-mail: tamburrini@unina.it

F. Santoianni (ed.), *The Concept of Time in Early Twentieth-Century Philosophy*,
Studies in Applied Philosophy, Epistemology and Rational Ethics 24,
DOI 10.1007/978-3-319-24895-0_32

Of special interest in computer science are prediction problems concerning the fulfillment of specifications on the design and implementation of computing systems. Specifications illuminate a chief difference between methodological problems arising in the natural sciences and those arising in both computer science and other technical disciplines. Specifications in computer science are imbued with human intentions and goals. Indeed, computing systems are usually designed and implemented as suitable means to achieve some given set of human goals and intentions. Therefore, specifications are *prescriptive* in ways that scientific laws and models are not, insofar as the latter do not necessarily involve the fulfillment of human desires and intentions.

1 Temporal Logic and Computing System Specifications

Major specifications concern temporal properties of computing system runs. Consider an algorithm controlling the traffic lights at the confluence of two roads. Algorithms of this sort are usually modeled as *reactive systems*, that is, as non-terminating processes modeled as sequences of states, where the transition from one state to another is determined by some given set of actions. Time is usually modeled as a discrete variable, and the sequence of time moments in each system run corresponds one-to-one to the sequence of its states.

A basic specification for algorithms controlling traffic lights at the confluence of two roads is that the two traffic lights should never be green at the same time. In other words, this is a "bad" event which should not be allowed to occur by any reasonable traffic control algorithm. Specifications for "bad" events that should never occur are usually called *safety* properties. Clearly, exceptions to some safety property occurred in one or more computing system run giving rise to the accidents that were mentioned above.

The safety property that the two traffic lights should never be green at the same time—like many other safety properties for computing systems—can be expressed in various propositional systems of temporal logic. One of these system is the Linear Temporal Logic LTL, which is based on the two unary propositional operators $<E>$ and $[A]$. The intended interpretation of the formula $<E>\ \Phi$ over discrete and linearly ordered moments of time is "Φ will be *E*ventually true at some future moment" and the intended interpretation of the formula $[A]\Phi$ is "Φ is true now and for *A*ll future moments."

By combining operators $<E>$ and $[A]$, one obtains derived temporal modalities. Thus, for example, the interpretation of the formula $[A]\ <E>\ \Phi$ is "always eventually Φ". In other words, at any moment i there is a moment $j > i$ at which Φ holds. Accordingly, the formal system LTL enables one to express future repetitions of events—not unlike some of the extended verbal tenses analyzed by Reichenbach, and amply illustrated by Malatesta with reference to both classical and modern languages.

Future repetitions of events play a crucial role in another class of specifications for reactive systems. Consider again the traffic light control system. Each traffic light satisfying the above safety property must additionally turn green over and over again—insofar no car incoming from one of the two merging roads should be left waiting forever at that junction. If one models the traffic light system as a non-terminating reactive system, this particular specification amounts to requiring that each traffic light at the junction should turn green *infinitely often* in the future. Statements expressing the infinite future recurrence of "good" or "desirable" events are called *liveness* properties of reactive systems. Reichenbach's pioneering analysis of extended verbal tenses revealed, among other things, the logical structure of statements expressing liveliness properties of reactive systems.

2 From the Logic of Specifications to the Methodology of Model-Based Predictions

Let me now turn to consider the role of specifications within what—following Reichenbach's well-known distinction between the context of discovery and the context of justification—one may aptly call the context of discovery in computer science. In particular, I will briefly comment on a formal method called *model checking* in theoretical computer science, pointing to its significance as a method for discovering new conjectural knowledge about target computing systems which is expressed by means of temporal logic formulas.

Model checking is a prominent model-based method which enables one to address a variety of problems from the following general class: "Does reactive system S satisfy the (safety or liveness) property expressed by means of temporal logic formula Φ?" Schematically, model checking involves three distinctive steps: the formal property specification step illustrated above; the building of a formal model M_s of the reactive system S of interest; the development and application of a decision procedure enabling one to establish whether Φ is satisfied by M_s. In other words, model checking enables one to decide whether, *according to* M_s, S meets or else violates Φ.

Models of reactive systems that one builds for the purpose of model checking some temporal property Φ usually take the form of *finite transition systems*. Transition systems are directed graphs whose nodes represent states of the reactive system, and whose edges represent actions enabling one to make transitions between states. To each state s of the transition system one associates a set of sentential letters that are true at s. A transition system is *finite* if the set of nodes, the set of transition actions, and the set of sentential letters are finite. Given sufficient computational resources, the model checking method enables one to apply an algorithmic decision procedure to a model of S as a finite transition system and a property specification given by some temporal formula Φ to determine whether property Φ holds in the model. The algorithmic decision procedure explores exhaustively the state-space trajectories in the model, and eventually returns the truth value "True" or the truth value "False".

One should be careful to note that the outcomes of the model checking decision procedure enable one to learn something about the behavior of the target reactive system *through the intermediary* of its representing model (transition system). The projection on the target system of judgments one endorses by reflecting on a model of the system only, is a form of model-based reasoning which is aptly called *surrogative* reasoning in the philosophy of science. By surrogative reasoning one draws new hypotheses on target systems that are solely based on model inspections. Hence, model checking enables one to discover regularities about computing systems, like those that are expressed by liveness temporal properties.

Hughes (in his "Models and representation", *Philosophy of Science* 64, 1997, pp. 325–336) distinguishes between three stages of surrogative reasoning: *denotation*, *demonstration*, and *interpretation*. In the denotation stage, one addresses the semantic issue of specifying in virtue of what models do represent some target system. In the demonstration stage, the target system recedes in the background, and the model's internal structure and dynamics come to the fore. Indeed, by inspecting the model's structure and dynamics, new significant behaviors of the model become apparent. Interestingly, this inspection process is thoroughly mechanized in model checking, insofar as it amounts to the application of an algorithmic decision procedure, which enables one to verify specifications expressing properties of trajectories in the state space of the model. Finally, in the crucial interpretation stage, each event or regularity identified in the inspection stage and concerning the model's behavior is turned into a conclusion about the target system.

To assess the significance of model checking for the context of discovery in computer science, one must carefully note that any conclusion about the target system that one infers in the interpretation stage *is a falsifiable hypothesis* that one may reject on the basis of empirical observations of the target system behavior. If one detects a discrepancy between actual system behaviors and the generalizations about the target system that are generated in the interpretation stage, then one achieves a falsification of the hypothesis which, in its turn, gives rise to a specific version of Duhem's disjunctive problem for model checking: in the face of an accepted falsification, one has to choose between model revision, specification revision, and verification procedure revision. In particular, *model* revision is required only if the above behavioral discrepancy is detected *and* one endorses the background assumptions that both specification and verification procedures are correct.

Model checking and other formal methods in computer science enable one to attain and test conjectural knowledge about computing systems by deploying methodologies that are on a par with scientific methodologies that are used in the natural sciences to address predictive problems about the behavior of systems which involve neither human design nor craftsmanship. Thus, for example, by verifying the correctness of some computer program P with respect to some specification R, one establishes the correctness of some model of P with respect to

R, and projects this result in the interpretation stage as a fallible hypothesis on the modeled program P. This view of model checking and its role in the context of scientific discovery is coherent with the broader claim that computer science is an empirical science. Computer scientists, like other scientists, introduce models of complex target systems, hypothesize regularities governing their behavior, and empirically test those regularities by experiment. There is no difference, from this perspective, between complex systems that are designed and implemented by man, on the one hand, and systems that develop without any human conscious intervention. Scientific knowledge about systems of both kinds unfolds through the discovery and the testing of empirical generalizations about their behaviors.

Editor Biography

Flavia Santoianni is Full Professor of Education at University of Naples Federico II. Director of the international open access journal RTH Research Trends in Humanities, she has published 26 books and several articles. Her books are translated into English and Spanish. Authoress of the Theory of Elementary Logic, her research interests concern bioeducational sciences, teaching and learning, and learning environments design.

F. Santoianni (ed.), *The Concept of Time in Early Twentieth-Century Philosophy*,
Studies in Applied Philosophy, Epistemology and Rational Ethics 24,
DOI 10.1007/978-3-319-24895-0

MIX
Papier aus verantwortungsvollen Quellen
Paper from responsible sources
FSC® C105338

If you have any concerns about our products,
you can contact us on
ProductSafety@springernature.com

In case Publisher is established outside the EU,
the EU authorized representative is:
Springer Nature Customer Service Center GmbH
Europaplatz 3, 69115 Heidelberg, Germany

Printed by Libri Plureos GmbH
in Hamburg, Germany